**WHOLE BEAST
BUTCHERY**

WHOLE BEAST
BUTCHERY

The Complete Visual Guide
to Beef, Lamb, and Pork

RYAN FARR

with Brigit Binns

Photographs by
Ed Anderson

CHRONICLE BOOKS
SAN FRANCISCO

DEDICATION

To my beautiful wife, Cesalee, and my son, Tanner. I can't imagine life without you.

Library of Congress Cataloging-in-Publication Data:
Farr, Ryan.
Whole beast butchery / by Ryan Farr, Brigit Binns.
p. cm.
Includes index.
ISBN 978-1-4521-0059-3
1. Meat cuts. 2. Meat cutting. 3. Cooking (Meat)
I. Binns, Brigit. II. Title.

TX373.F37 2012
641.6'6—dc22
2011002619

Manufactured in China

MIX
Paper from responsible sources
FSC® C008047

Designed by Vanessa Dina
The author has no affiliation with any of the entities on pages 237 and 238 and is not paid to recommend or endorse the resources and businesses listed.

10 9 8 7 6 5

Chronicle Books LLC
680 Second Street
San Francisco, California 94107
www.chroniclebooks.com

Acknowledgments

Attempting to thank on a couple of pages family, friends, and colleagues who have helped me in my life is an impossible task. I am very fortunate for every laugh, tear, and piece of bacon we have shared together. Thank you all for your support, guidance, and hungry bellies.

Cesalee, you are an amazing mother, partner, and loving wife. I am very proud of everything we have created together and excited for what tomorrow brings.

Mom and Dad (Hulyn and Ron Farr), thank you for raising me with unconditional love and for your unwavering belief in me. (I do know I was a little shit.) I look forward to raising my son with the same compassion, support, and love that you gave me.

Lauren (Lauren Farr), you have been my partner in many crimes and many BBQ meals in KC. I am proud to be your big brother. Bonapatche!

Mema and Pop (Carolyn and Hughie Habighorst), waking up before the sun rose to get a jump-start on breakfast is one of my favorite memories. My love for food began with the amazing people we met, the restaurants and diners where we ate, and the places we traveled.

Grandma (Guyneth Farr), as a child enjoying your cherry pies, buttery toast, and perfectly cooked eggs created memories of simple perfections. Those moments were realizations that food is much more than just substance. Thank you for showing me less is more. And a deep thanks for not locking me out of the house when I let all the rabbits free from their pens.

Autrey, Hogan, and Leffler families, thank you for your love and your welcoming arms, for embracing us into your family. The endless good times and wonderful Southern hospitality are very dear to me.

Devin Autrey, you are truly an inspiration as a friend, man, and chef. You always followed your heart and your stomach; your life has truly been a model for so many.

Cole Mayfield, you are a very good friend and a talented chef. Your approach to life and food with the utmost passion and love is commendable. Your meticulous recipe testing and tolerance of my BS for all these years have been invaluable. Thank you.

4505 employees past and present, your support, dedication, and continual enthusiasm have helped us find our place in the culinary world. Thank you immensely.

Marc Leffler, you are a funny and talented man. Thank you for your one-liners, laughter, and endless help building the 4505 brand.

C.H.E.F.S. staff, all of you and all the students helped me find my love for teaching and gave me the opportunity to build our company. Thank you.

Huge thanks to all the ranchers and their families for the endless work raising amazing animals. Mac Magruder, Don Watson, and Lee Hudson, thank you for the beautiful steers, lambs, and pigs—without your wonderful animals, none of this would be possible.

Lulu and Dexter and all the amazing staff, farmers, and volunteers at the CUESA Ferry Plaza Farmers' Market, your dedication to the community and to providing for others is an inspiration (and you are all great friends, as well).

Kent Schoberle, you do great work, sir. Thank you for your continuous dedication.

Thank you, to the talented staff of Chronicle Books: Vanessa Dina, your dedication and ability to see the beauty of butchery while keeping the functionality clear in the design is sheer talent. Lorena Jones, thank you for humoring me and believing in me. Your determination and boldness are paving the way and setting an example for so many.

Brigit! Your hard work and love for good food has elevated our professional relationship into a friendship that will last forever.

Ed Anderson, your ability to capture the beauty of every cut and technique has added an element to every picture that is rarely captured. You truly have skills, sir. Thank you.

Thank you Carole Bidnick for your endless work making all this happen.

To all the film and production crew, it was a pleasure working together. Thank you.

Thank you to the Bay Area! Your beauty, bounty, and personality are something very special and are the key ingredients in 4505 Meats.

To everybody who has supported me since day one: I owe you the world, and I love you all!

TABLE *of* CONTENTS

CHAPTER ONE

CHAPTER TWO

CHAPTER THREE

INTRODUCTION

I'm excited to share my love of butchering and passion for whole-animal utilization. With this book you'll learn how to break down a whole animal and utilize the *entire* carcass. I am going to cover the basic cuts and explain everything you need to get started.

The great thing about whole-animal utilization is that there are many different ways to approach breaking down the animal. In *Whole Beast Butchery*, I show you some of my favorite techniques. Because I'm a classically trained chef, my view of the animal is based on the end result. I look at how every single piece of the animal can be used and consumed without any waste. That's why my approach is sometimes a little unconventional; it's my own point of view, and it reflects how I developed my style. Once you have learned the basics, you can develop your own. My tutorials will help you reach that point by showing easy ways to break down a whole animal or a section, whether you're a home cook or a professional working in a restaurant.

You can use all these techniques as a foundation for doing your own butchering, but don't think that means you can replace your local butcher. Butchery and whole-animal utilization are arts that are often handed down from generation to generation and take years to master. Throughout my career, I've been inspired by the men and women who have chosen this honorable profession. I would never suggest that we don't need butchers—hey, we need a lot *more* butcher shops and opportunities for butchers!—but by using this book, you can do some of what they do.

We also need to support our local farmers and ranchers. There are probably more than a few sources of grass-fed animals in your hometown, but it might take some digging to find the family-run ranches that produce them. Do the digging.

Start by chatting with your local butchers. Go to the farmers' markets and delicatessens. Find these farmers and ranchers. Talk to them and make friends. Buy local, and help small businesses and family farms grow. This is what it's all about.

Why Butcher Your Own Meat?

Butchering in this country has broken out of the restaurant kitchen and moved into the home. This trend is a result of several major changes in our society, including a stronger desire to live up to higher moral and ethical standards, concerns for the safety of our food supply, and a renewed spirit of self-reliance. Now that the "givens" of our daily lives are in jeopardy—our economic security, the dependability of our food supply, and the safety of our generous natural resources—raising our own food affords us both a sense of security and a tremendous feeling of accomplishment. Even if we can't raise our own animals, many of us want to know how the animals we consume were raised and what they were fed.

Home butchering is the logical next step for those who raise their own vegetables and chickens, preserve the bounty of the land and field for off-season meals, and care deeply about what they feed themselves and their families. When you decide to butcher a whole animal or a part of one by yourself, as I hope you will, you are almost always going to be buying that animal locally. By doing so, you are supporting a local business as well as your community: If you buy a pig raised by your neighbor on his farm or from a rancher outside of town, he will have more money to spend, some of which might even come back to you. One thing is certain: That farmer or rancher can now raise more animals, raise them well, and supply more concerned consumers with better-quality meat—meat they can trust as well as enjoy.

Where Did It Come From? What Did It Eat? How Well Did It Live?

These are important questions that supermarket shoppers don't get to ask. The small-scale farmers and ranchers with whom you will most likely be striking a deal have herds that are smaller than those of the national meat producers, which means more attention and better conditions for the animals during their lifetimes, as well as a more humane end. If an animal sacrifices its life for our table, we owe it this respect. Vote with your wallet! Many of us who eat meat have become disillusioned with the factory-farming process in our country, yet we also intend to keep eating meat—only now we want to eat meat responsibly and fully understand what we are consuming. To this end, *Whole Beast Butchery* gives you the knowledge you need to exercise a degree of control over your own nourishment, beyond the supermarket.

Uncommon Cuts

Because of the industrial nature of our national meat-processing industry, many of the cuts of meat—like neck, trotters, offal—that were familiar to our ancestors, who lived closer to the land, are no longer staples on our tables. We have grown lazy, used to buying Styrofoam packs of rib eye steaks, unaware of the many wonderful parts of the animal that suddenly become available for experimentation and hearty enjoyment when you butcher at home.

It's All Edible

Cutting up a whole animal may seem a little daunting at first, so to help you conquer your fear, this book will provide you with step-by-step instructions, as well as plenty of visual aids. Remember, there's no such thing as a mistake and no set rules of butchery, except for what your stomach says.

The wonderful thing about the animals we eat is that virtually everything is edible. If you make a cut in the wrong place or your roast looks a little sloppy, no worries; it'll still taste great. So, cut with confidence, plan your next move, and make straight cuts. As you come to understand the musculature of each animal and the characteristics of each of its parts (Is it lean or well marbled? Is it a hard-working muscle that requires slow-and-low cooking to retain tenderness?), you will come to intuit the correct method for cooking your "mistakes" and the various off-cuts that might have gone to waste. That's what whole-animal utilization is all about.

A Whole Lot of Meat

Depending on the size of the animal and which animal you choose, you are going to end up with anywhere from 40 pounds/18 kilograms of meat to more, much more. A whole, or even half, a grass-fed steer is a very large amount of meat. In most cases, it is far more than one family can eat or store—about 700 pounds'/317 kilograms' worth. So I recommend that you plan ahead when you set out to purchase a whole carcass or even large primal cuts. In fact, you may want to start with a primal (say, on a steer, one of the forequarters; or on a lamb, the saddle section). After you are comfortable with the art of breaking down a primal cut, you will be ready to move on to a whole lamb or pig, or half a steer. In any event, be sure to plan how you will store the meat before you begin sharpening your knife.

Freezing the Meat

I am not a huge fan of freezing meat, but for most home cooks it is unavoidable. That said, you can take steps that will help keep your meat in optimal condition once you consider several factors, one of which is fat content. Fat contains no water, so during the freezing process there is less crystal-lization in fatty meat than in lean meat. Therefore, I advise you to eat the lean cuts first, grilling or roasting up your chops, loin, and tenderloin, and to freeze the chuck, shoulder, belly, and other fattier cuts, plus the off-cuts that you will eventually turn into ground meat and/or sausage. (Don't grind the meat before freezing, as the thawing process will yield dry meat. Instead, freeze it in large pieces. When you're ready to make burgers, defrost the meat gently, cut it into chunks just slightly smaller than the opening of your grinder, and *then* grind it.) Another factor is size: bigger pieces of meat survive freezing better than smaller pieces. A pork shoulder will freeze and defrost in far better condition than a pork chop. But, in some cases, you will still have more meat than you and your family can eat before it begins to go bad. If you need to preserve a large, lean cut such as a leg, I heartily advise you to cure it instead. For a leg of beef, make *bresaola*. For a leg of pork (a.k.a. "fresh," or "green," ham), cure it first, and then either smoke or dry it by hanging—then make your own prosciutto, for instance.

Be aware that in the process of freezing and cooking, meat will lose from 10 to 20 percent of its weight. This weight loss, which is a combination of water and fat, is the result of several factors: freezing and defrosting, the rendering of fat, and the cooking process. So if your goal for a week-night dinner for four people is 5 or 6 ounces/140 to 170 grams of meat per person, you should freeze the meat in 2-pound/1 kilogram portions. This could be four rib eye steaks or enough chuck for four hamburgers. In any event, please don't freeze meat for more than three to four months, or it will turn into something other than the beautiful animal you started with.

It Takes Four Families

One approach that many home cooks are taking is partnering with several other families to purchase and butcher a whole animal or, in the case of a steer, half an animal. Because of the way animals are structured, I find that a division into four por-tions makes sense, as each family ends up with a more manageable amount of meat to eat and store. Once you find three other like-minded families, buy your animal and begin the process: Make sure that you have adequate freezer and refrigerator space, plenty of butcher paper, markers for labeling, a scale, and plenty of clean cutting boards and sharp knives. For a steer, each family will take home between 150 and 200 pounds/68 and 90 kilograms of meat. Well managed, this amount of meat can feed a family for a long time!

Another Way to Approach Breaking Down a Whole Animal

If you aren't quite ready to take the plunge into carving up a whole animal, there's another option (but you will still need to understand the structure of the animal and all the characteristics of the meat from its different parts). Many communities—especially those in areas where there is a lot of hunting—have a game processor/private slaughter-house. Hunters bring the deer they get to these processors, who break down the animal into portion sizes and roasts per the hunters' choice, often turning much of the animal into sausage prepared according to the processor's own time-honored recipe. These folks are a fine resource for the family who would like to break gently into obtain-ing local meat. You could buy the animal from a local rancher or farmer, and then have the butcher process it for you. For most people, this is still a far better choice than buying meat from the super-market. Remember, however, that you will have to tell the butcher how to process and portion the animal, which means familiarizing yourself with all the options for the animal you've chosen, plus taking into account the size of your family and your favorite menus—the same decisions you'll need to make if and when you butcher your own animal. Or there's another option: have the butcher break down the animal into "primals" (the large sections of an animal). Then you can freeze the meat in primal cuts, allowing you to take your time defrosting and breaking down each primal into subprimal and individual cuts.

The Choice Is Yours

The beauty of butchering the whole animal is that you are the boss when it comes to breaking it down, but you will need to understand all the different options in order to make the best decision based on your needs. Not every cut of meat with which you are familiar can physically come from the same animal. The animal only has a certain number of ribs, for instance. If you want tenderloin medallions or filet mignon, you won't be able to cut porterhouse or T-bones from the same side of the animal.

For a big party, you might decide to buy and break down a primal cut that you can then cook in several different ways. For instance, a forequarter of beef can be broken down into seven or eight flintstone chops, a large shank for braising, and plenty of meat for grinding into meatballs or sausages. On the other hand, you might decide to break down a forequarter into ribs for Texas-style smoked beef ribs, a bone-in prime rib for a roast, and a brisket, which you could brine and smoke.

Similarly, a center-cut loin section of pork could be portioned according to one of the two following schemes, but not both:

SCENARIO ONE
Frenched and boneless pork chops, plus a whole belly for curing and smoking.
or
SCENARIO TWO
Ribs for smoking, plus a loin for slow roasting and a belly for roasting, crispy-skin style.

For a lamb, which is much smaller than a steer, you could choose between the following schemes for a primal cut, called a whole saddle:

SCENARIO ONE
Remove the loin bone and roll up both sides toward the center, then tie it up tightly for a large and impressive "saddle roast."
or
SCENARIO TWO
Remove the belly for grilling until crisp and cara-melized, plus cut the loin and tenderloin, with their bones, into six porterhouse T-bone steaks.

Let's Get Started

With this book, you have the tools you need to break down and utilize whole animals. Your family will be eating better meat; you'll be fully using all the nutritious meat on the fantastic animal you've bought; and you'll know everything there is to know about the meat you put into your body.

A Butchery Primer

The craft of butchery is equal parts art and science—with, sometimes, a morsel of engineering. At times, you will follow your instinct; at others, an anatomical chart will be indispensable. Arm yourself with in-depth knowledge and the best tools, and you will be successful at this ancient (now modern) craft.

Basic Guidelines
There are no set rules (except to follow your stomach!), but these are the guidelines I always follow:

1. Hatchet, saw, cleaver for bone; knife for flesh and skin.

2. Leave on as much fat as possible; you can always trim it after cooking. I don't like to trim off any flavor (fat) before cooking.

3. The bottom line is that this is food. If you make the wrong cut, it's no big deal; it's all edible and still great tasting. You can eat everything but the oink.

4. Always save the bones and trimmings for stock; waste nothing from these fine animals.

5. The worst thing in any kitchen is an insecure or shaky knife. If you're going to cut, just go at it with confidence. Be sure to keep all your body parts—fingers, hands, arms, and legs—away from the sharp cutting instrument.

6. In the how-to pictures in this book, I tie up the roast right away to show you the technique, but in reality I always season before tying up a roast, and so should you.

7. Salt is a tool; learn to use it. You can cut up the animal perfectly, but if you don't know how to salt, it won't be delicious.

8. Ideally, cook in a convection oven; it'll give you the crispiest skin.

Tools
I bought my **hatchet** in Jackson, Mississippi, while visiting friends and getting a good dose of the South. I needed something that would enable me to get into tight places but still apply a lot of force. A cleaver has three times the amount of blade surface, so you can't really get it into tight spots. Also, when you're in tight spots, you usually can't come down with a lot of force. The hatchet has a lot of weight, so I don't have to really whack it hard. I just tap, and it goes through a lot of the softer bones beautifully. I wouldn't use it to go through harder bones like shanks or legs, though. It's perfect for separating the bottom of ribs, and the soft metal is easy to sharpen. Safety tip: Hold your other hand behind your back before you come down with the hatchet to go through a bone or a joint.

I use a **cleaver** when I want a longer blade surface for chopping bones for stock, and for bigger things that I want to cut or remove.

My **butcher's knife**, which is nice and long and curved up like a saber, is good for long cuts into the meat and for cutting final portions. Curve down with the cut and then curve up at the end, almost with a rolling action, so that you don't end up making jagged cuts.

I use a **boning knife** pretty much for everything when I'm butchering meat because I've been using it for so long that it's like an extension of my hand. There might be other tools that would work as well, but this is what I learned on. It's great for working around the bone and for following the sinew, as well as for working in other tight places, whereas the butcher's knife is better for longer cuts, like portioning servings.

I couldn't function without a **bone saw**. I have a band saw, but I don't use it that often. The saw is used to cut through bones and pieces of cartilage. I use it to go through the backbone either horizontally or vertically and for places where the hatchet won't work.

On the bigger animals, especially steers, whenever I finish sawing, I use a **bone scraper** to get rid of bone dust. (On smaller animals like pig or lamb, I just use the back of my knife or a damp towel.)

I use a **bench scraper** for scraping down my board when working with beef. Beef fat, which is sticky and gummy, can be tough to remove with just a dry towel. To keep the board nice and clean, use the scraper.

My **meat hook** helps me get a good grip on the animal when I'm moving it. The animal is slippery, wet, and fatty, especially when in direct contact with my warm hands.

A **rubber mallet** is extremely helpful when making some of the smaller cuts through the bone, like we do with lamb chops, when we don't want to use the saw because the friction would damage the delicate flesh. We can tap the cleaver through with precision, whereas if we were to use the cleaver freehand, we wouldn't have as much control.

The **trolley hook** (or meat hook on wheels) enables us to hang the animals up on the meat rail, so they are not resting on a shelf. This gives us better air circulation, with no chance of moisture buildup. You can spin the animal 360 degrees, and you can hang a bigger animal and let gravity help you out as you butcher.

I use a **trussing needle**—mine is called a roast-beef needle—for penetrating tough, thick surfaces like pig skin. The handle lets me push as hard as I want to. You can always poke through a small hole with a knife, but this tool makes the job much easier.

Heavy-gauge, all-cotton **butcher's twine** helps hold everything together nice and tight. We can use it to tie up roasts that would otherwise be floppy and not cook evenly.

There are two reasons it's important to have a good **scale**. First, when you follow my ratios and percentages in the recipe formulas, you will have to weigh your ingredients in order to maintain the consistency of the ratios. Also—especially for restaurant chefs, as well as home cooks—you *must* weigh portions as you cut, in order to maximize profitability and consistency.

I'll never work in a kitchen without a **mortar and pestle**. With this, I can take my whole freshly toasted spices and break them down for rubs. This tool opens up all the flavors.

Salt is a *really* important tool. Of course, just like all tools, you have to know how to use it. A lot of recipes tell you to use just a little bit, but that's not going to cut it. You can butcher the meat perfectly, but it won't taste good unless you season it. Whether you are curing or cooking, learn as you go and keep notes, because you probably won't remember how much salt you used.

I always have both an external and an internal **thermometer**. I use the external to check the temperature of the oven or fryer when I am about to cook. An internal, probe thermometer, ideally one with an alarm, is crucial to avoid overcooking any meat.

You'll also need a **calculator** to make sure the ratios are correct when you scale the recipes up or down.

Techniques

Owning the right tools is only a part of learning the craft of butchery. Knowing how to use and care for them is the next crucial step.

Sharpening

It's important to keep your blade nice and sharp all the time, because if you lose your edge it's hard to get it back. You have to stay on top of it. I sharpen my knives on a Japanese whetstone every single time I use them. On a daily basis, I use the 1,000-grit stone. If you lose the edge, you've got to return to a coarser grit, like 700. I use the 1,000 stone on my butchering knives, but if I want to sharpen one of my softer Japanese blades, I use a 10,000-grit stone. (The reason for this is that the blend of steel in the Japanese knives is more delicate than that of the steel in a butcher's knife; if I used a coarser grit, it would make the softer knives wear down to nothing way too soon.)

Press down hard when you're sharpening. (On my Japanese knives, I don't press so hard.) Start with seven times up and down, then flip. Another seven times, then flip again. Next, do six times, and flip. Six times, then flip. Then back to the other side, and do five times. I'll keep going down to one only if I think the edge is not sharp enough yet.

I use a **diamond steel** to hone the knives that have already been sharpened on the Japanese whetstone. I only do it two or three times, whenever I feel I've lost the sharp edge. If I lose the edge, I have to go back and resharpen the knife on the stone.

Keeping It Clean

When I'm working, a **damp cloth** is probably more important than my knife, hatchet, or cleaver. After *every* single step, you need to wipe off your board *and* your knife thoroughly. Every time. Otherwise, fat will build up, the board will get sticky and gummy, and you won't be able to work. When the cloth gets dirty, just rinse it off, squeeze, and keep going. A dry towel just pushes the fat and protein around. If you're not working clean, there's no point in working.

Working Close to the Bone

Whenever you are removing bones from flesh, keep the knife on the bone at all times. This way, you will be able to remove the bone with as little meat still attached to it as possible. Be sure to leave the maximum amount of meat behind on the roast or steak you are preparing. Always use the bones and trimmings for stock, so nothing will be wasted. But keep in mind that clean bones are the sign of a pro. Your mantra should be: highest yield, minimal usable meat wasted.

Whole-Animal Utilization

Whole-animal utilization is not just about using all the parts of the animal—including the offal, the lesser-known cuts, and organs—it's also about making sure there are no scraps left behind, which is also a great way to get the most value from your whole animal. Use the best scraps to make sausage and other scraps to make stock. Then poach your sausage in the stock. Then reduce the stock and make a sauce. You can also use the trimmed bits to fortify your stock so that you can use it to make a beautiful sauce after you've used it for poaching.

> **"The head and 'extras' are my favorite parts of any animal. The options are endless: There's a lot of skin, gelatin, bones, fat, and thus flavor—items you can't really find in a butcher shop. In the past twenty or thirty years, these cuts have become uncommon in meat cases."**

Making Stock

Although a whole book could be written about stock, stocks are actually really easy to make, and they add fantastic complexity to cooking.

Simply cover your bones and vegetables (in the form of a mirepoix, a mixture of celery, onion, and carrot in equal parts) with water, bring it to a simmer, and keep it at a very gentle simmer for at least 5 and up to as many as 24 hours, adding water as it reduces. You should be able to really smell and taste the meat, whether it's beef, pork, lamb, poultry, or game.

FOR DARK STOCK: Roast the bones and mirepoix until golden; add tomato paste for a little sweetness to counter the slight bitterness from the browned vegetables.

FOR LIGHT STOCK: Don't roast the bones or mirepoix before adding water, and add a little dry white wine.

Cooking

I like to cook with strong flavors on a nice grill over an open fire. Meat is rich enough on its own, so I would rather add bright flavors, such as vinaigrettes, fresh herbs, and vegetables, than a rich sauce. Keep it bold and make your palate dance with flavor.

I use a lot of fresh herbs in my braising. I don't mind if the leaves stay behind but I don't want the thick stems, so I tie the stems with a piece of string. Then, I can just pull off the bare, tied stems.

In the step-by-step photos, I proceed straight to tying the roasts after they've been boned. If I were preparing them for cooking (not demonstrating how to butcher and tie them up), I would add plenty of seasoning *before* I began to tie them.

> **"Fat is good."**

With lean meats especially, manage your temperature and *don't rush it.* Be sure to let your meats rest correctly, covered in foil or in a warm area or even in fat or butter.

How to Use the Recipe Formulas

In nature, there are many variables: The size and age of the animal may range widely. One lamb shoulder may weigh 4½ pounds/2 kilograms, and another one might weigh 8 pounds/3½ kilograms. If the butcher gives you 10¾ pounds/5 kilograms of brisket, and the recipe only calls for 10 pounds/4.5 kilograms, you'd be wasting ¾ pound/340 grams of protein. In the interest of preventing waste and making these recipes amazingly user-friendly, I've made them completely adaptable, based on the weight of the meat being prepared. So, instead of wasting meat or spending hours with a calculator trying to adapt a recipe, all you have to do to maintain the consistency of the dish is multiply the percentage of the ingredient by the total weight of the recipe in grams. This not only prevents waste, thereby saving money, but also helps ensure the meat will consistently turn out terrifically, no matter how much you start with. This formula is always based on the yield of the recipe in grams. The process is far more intuitive when working in grams, but we've provided ounces and, in most cases, volume measurements (cups and tablespoons), as well as the percentages. Ideally, though, use a scale and a calculator when following these recipes.

DESIRED WEIGHT OF RECIPE IN GRAMS X % OF INGREDIENT = WEIGHT OF INGREDIENT

EXAMPLE:
Recipe yields 1,000 grams X 10% salt = 100 grams of salt

(Remember: to convert a percentage into a decimal, shift the decimal point two spots to the left [10% = .10]. This will make your calculations much easier.)

> "To maintain consistency when scaling a recipe, a good calculator is more important than a good knife."

Brining

I am a big fan of brining and corning. If I have time, I will brine anything and everything; it really makes a huge difference. Brining makes the flavors blossom and helps retain the juices in cuts that would otherwise be dry. When in doubt, brine, brine, brine! There are two key variables to understand about brining: one is the ratio of salt and sugar to water, and the other is the length of time the protein spends in the brine. For a general rule of thumb, brine a piece of meat that is 6 inches/ 15 centimeters thick for 3 days, 3 inches/

7.5 centimeters thick for 36 hours, and 1 inch/ 2.5 centimenters thick for 12 hours. Always document the amount of time that you brine, as well as the results. That will help you maintain consistency and improve your results going forward.

Master Brine
YIELD: 4.73 liters/1 gallon and 1 quart

This recipe is a starting point, but there are many possible variations. If you're not a fan of hot flavors, go ahead and omit the chiles. Always use a tall, narrow nonreactive container only just large enough to hold the protein, so the brine will go as far up as possible. The brine must cover the protein *completely*, so scale the quantities here up or down as necessary.

granulated sugar	2 cups	13.6 oz	385 g	8.5%
kosher salt	2 ½ cups	20.4 oz	578 g	12.7%
whole black peppercorns	¼ cup	1.2 oz	34 g	0.7%
whole coriander seeds	6 tbsp	0.8 oz	24 g	5%
dried bird's-eye chile or Thai chile	3 small	6 oz	17 g	0.4%
water	16 cups	123 oz	3500 g	77.1%

Combine everything in a large pot and bring to a boil. Once the sugar and salt have dissolved, remove from the heat. Transfer to a tall nonreactive container that will fit in your refrigerator and let it sit uncovered to cool. When the brine is at room temperature, refrigerate until it is completely cold. Add the meat, and brine as directed.

General Sausage-Making Tips

When making sausage, the meat must stay at or below 45°F/27°C at all times during the process. The ideal temperature is 38°F/20°C. Before cubing the meat and then again before grinding it, "open-freeze" it by placing the meat, uncovered, in the freezer for 30 to 60 minutes or just until the surface has a little crunch to it. The intention is not to freeze the meat solid but to get the exterior hard enough so that it's brittle on the outside but still soft in the middle. When open-freezing, it's not necessary to distribute the meat so it's not touching other pieces; the meat on the top will freeze first, and that will be enough to lower the temperature for the whole batch.

When you are not using them, keep all the rest of your ingredients and equipment, including the grinding equipment and the stuffer, cold by storing them in the refrigerator.

Cut the meat into squares that are slightly smaller than the opening in the meat grinder. Do not force the meat into the grinder; it breaks down the delicate cells and warms up the fat before you are ready for that to happen. If you let the auger gently "grab" the meat and move it toward the blade, you will get nice, clean cuts without overly compressing the meat.

I recommend that you do not use the sausage stuffer that comes with stand mixers. With these machines, you have to push too hard to move the meat mixture toward the horn, pulverizing the meat and thus wasting all your efforts to keep it from being compressed during the grinding and mixing process. I suggest you invest in a 5-pound/2.2 kilogram vertical, crank-operated manual sausage stuffer, which is widely available on the Internet.

A lot of recipes tell you to prick a sausage before cooking it, but I'm not a big fan of this technique—you lose delicious juices. Only prick where you see visible air pockets. If you cook the sausage fast over high heat, it will probably burst; so don't do that! Cook it slowly and gently—and only prick the sausage if you see airholes. By cooking slowly, the skin will expand, but not split, and all the juices stay in the sausage.

Sausage casings are sold in large quantities—units of measurement called *hanks*—which can be ordered online or from a specialty butcher. One hank will stuff a minimum of 75 pounds/34 kilograms of sausage, but you can pack the unused casings in salt and freeze or vacuum-seal them for your next batch.

CHAPTER ONE

BEEF

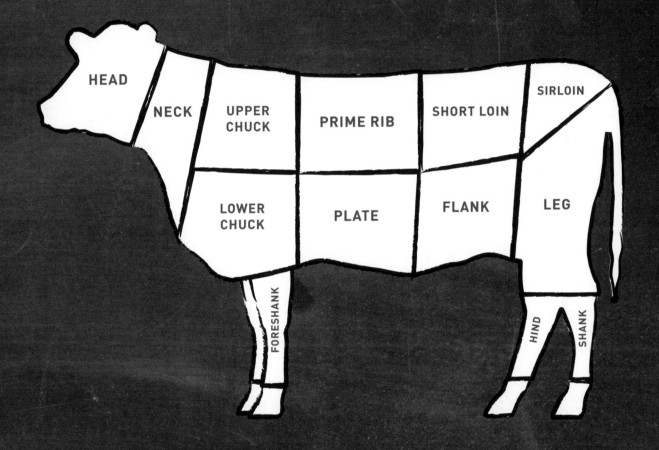

In this chapter, we will be working with a beautiful grass-fed steer that has been aged for ten days. All mammals have the same basic skeletal structure—from pigs to rabbits to cows—so we break them down in a similar way. But, this is a *very* large animal. So, the first step with a steer is to break it down into four quarters.

To me, the **neck** is the trophy cut of the whole animal! It's full of flavor but rarely found in a butcher's case. It's a small muscle with a good amount of fat all the way through. It gets a lot of exercise, and that's why it's so flavorful. This is a delicate piece of meat, but it's still tough enough to require braising.

The **skirt** is the cut I always use first on the forequarter, because it has been exposed to the elements more than other parts. It's one of my favorites—along with the hanger—and it's what we have for lunch when we're butchering beef. Skirt and other so-called flap meats *must* be sliced across the grain; otherwise, they'll be tough and stringy.

The **flintstone rack** is a long, bone-in prime rib with the short-rib meat left on. A flintstone chop is the same thing but with only one rib bone. The short-rib meat isn't super-tender but has a ton of flavor. The flintstone is a cut that you're unlikely to find unless you cut it yourself—or you're willing to pay an arm and a leg for one chop!

All of the **chuck meat**, on the lower part of the forequarter, is great for smoking or for making into burgers or sausage. You can do almost anything with it, because it has great fat. I like to stud it with garlic cloves, rub it with seasonings like I would a pork shoulder, and smoke it.

Everything in the lower section of the forequarter is really tasty but needs to be cooked carefully. **Brisket** can be buttery-soft or tough as an old boot; you just have to know how to treat it. I always make sure it's brined or cured before cooking, and then I cook it slow-and-low. It also makes excellent meat for burgers.

I like using the tough meat of the **breast** and **plate** to make hot dogs. When I puree and pulverize the meat, it doesn't matter that it's tough. With a leaner, high-protein cut such as this, I am able to add more fat in the emulsifying process. The breast is also great for confit, especially with aged beef. But keep in mind that this is very lean meat. You could also make this meat into beef stew or use it for tacos and chili.

The **short ribs** are the extension of the prime rib bones. The way I like to cut them yields more than just short ribs; it gives you the upper plate meat as well. There's a lot of good plate meat still on there, so it's a good idea to cook it in the same way as the short ribs.

Flank is a popular cut that a lot of people are familiar with, but all of the other **flap meat** in this section is just as tasty. Flap is one of those hard-to-find butcher's secrets. A lot of good meat gets left behind in the standard meat-packing process that isn't marketed; flap is just one great example of that. In general, you'll only get flap meat if you are buying a quarter of beef or shopping at an artisan butcher shop. Flap meat is best cooked quickly over high heat.

The **T-bone** and **porterhouse** are where it's at, steak-wise. Even though the meat is leaner than its neighbor, the rib eye, it's surrounded by a lot of nice fat. I grill it over high heat, starting out with direct heat over hardwood coals, so I get good caramelization on the fat, and then I finish it over indirect heat. When it's almost done cooking, stand it up on the grill, bone-side down, so the bone heats up and transfers heat into the meat, adding the *great* flavor that comes off that bone.

The **spider** is a small cut on the inside of the hip bone right above the joint, and because it's such a great snack, it usually doesn't make it farther than the butcher's stomach. I sear it, flip it, and then finish cooking over gentle heat. You don't need to add a lot of fat because all of its own fat renders out and caramelizes.

The **round** is a challenging area for everybody who cuts and cooks beef, because it's very lean—which means it has the potential to be tough and dry. It's not as forgiving as other cuts that have a lot of fat in them. This is where I make my cowboy steaks: I square off the round, lard the whole thing with lardo (cured pork fat) or fatback, and then make big spiral-cut steaks. I rub them down with spices and herbs and baste them while they cook with butter and my chosen flavorings— such as rosemary, thyme, roasted garlic, shallots, and olive oil. The fat drips off and produces nice smoke. I don't mind a tougher cut as long as it has great flavor.

Don't be too eager to trim off the **fat** from these luxurious—but lean—cuts. Fat is where the flavor lives! Let's repeat that: *fat is where the flavor lives!* However, there is a lot of extra fat in the hindquarter section, including all the hard fat on the short loin. This is great for rendering down; you can fry chicken or potatoes in it—you can even use it for a crust for a potpie. Incredibly tasty stuff.

Butchering and fully utilizing a whole steer is challenging and very physical; once you have finished a long day of cutting, reward yourself: grill up a tasty, juicy piece over a hardwood open fire and enjoy the butcher's ultimate reward. You've earned it!

FOREQUARTER, STEP 1: **Whole forequarter standing up on its side.**

FOREQUARTER, STEP 2: **Begin breaking down the forequarter. Make the first cut between the 5th and 6th rib bones. Cut through all the flesh with the knife before sawing.**

FOREQUARTER, STEP 3: Saw through the bone on the loin side first and then saw through the cartilage of the breast side.

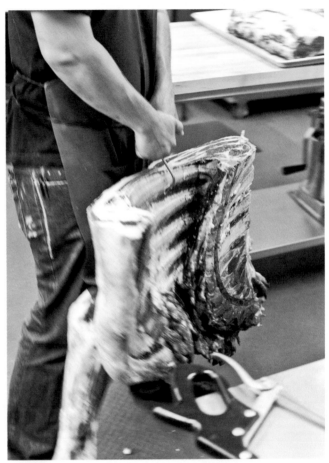

FOREQUARTER, STEP 4: This cut will release the entire chuck from the rib. This piece of chuck will be hung in the walk-in for 14 days and then made into dry-aged burgers.

FOREQUARTER, SKIRT, STEP 1: Now, with the rib section that we left hanging, remove the skirt from the inner side of the forequarter.

FOREQUARTER, SKIRT, STEP 2: Continue pulling off the outside and inside skirt, by making small cuts.

FOREQUARTER, SKIRT, STEP 3: Trim up the outer, dried, and oxidized parts of both skirts.

FOREQUARTER, SKIRT, STEP 4: Pull off the fat, beginning at the head of the skirt.

FOREQUARTER, SKIRT, STEP 5: Trim off the fat with the knife.

FOREQUARTER, SKIRT, STEP 6: Pull the silver skin off the skirt.

FOREQUARTER, SKIRT, STEP 7: The finished inside skirt section, folded over.

FOREQUARTER, FLINTSTONE RACK, STEP 1: Saw through the feather bones on the back side of the loin to allow you access with the knife.

FOREQUARTER, FLINTSTONE RACK, STEP 2: Saw down to remove the chine bone.

FOREQUARTER, FLINTSTONE RACK, STEP 3: Steady the rib section with your arm as you saw down all the way to the bottom.

FOREQUARTER, FLINTSTONE RACK, STEP 4: Flip the rib section around and come back with the knife to peel off the feather bones.

FOREQUARTER, FLINTSTONE RACK, STEP 5: Cut down along the inside of the feather bones.

FOREQUARTER, FLINTSTONE RACK, STEP 6: Mark the first cut on the whole rib section right at the beginning of the cartilage at the tip of the rib.

FOREQUARTER, FLINTSTONE RACK, STEP 7: Mark the cut on the chuck side at the tip of the rib as well.

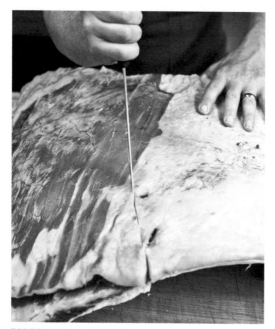

FOREQUARTER, FLINTSTONE RACK, STEP 8: Flip over the rib section and connect the two marks, cutting through the meat to the ribs.

FOREQUARTER, FLINTSTONE RACK, STEP 9: Saw through the ribs, using the marks as a guide.

FOREQUARTER, FLINTSTONE RACK, STEP 10: The whole flintstone rack, with the plate and the tip of the shoulder blade, plate meat left on.

FOREQUARTER, FLINTSTONE CHOPS, STEP 1: Pull off the plate meat.

FOREQUARTER, FLINTSTONE CHOPS, STEP 2: The flintstone rack from the rib side.

FOREQUARTER, FLINTSTONE CHOPS, STEP 3: Make an incision exactly halfway between each rib to create chops 3 to 4 inches/7.5 to 10 centimeters thick.

FOREQUARTER, FLINTSTONE CHOPS, STEP 4: Flintstone chops: a gift of love.

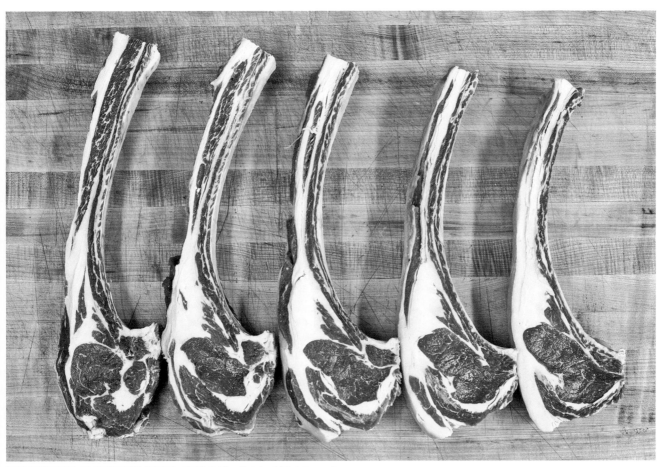

FOREQUARTER, FLINTSTONE CHOPS, STEP 5: **The finished flintstone chops.**

FOREQUARTER, SHANK, STEP 1: Now, working on the *other* forequarter, let's lay this on the cutting board and begin breaking it down. First, locate the joint between the shin and the shank.

FOREQUARTER, SHANK, STEP 2: Begin cutting through the flesh between the shank from the arm bone.

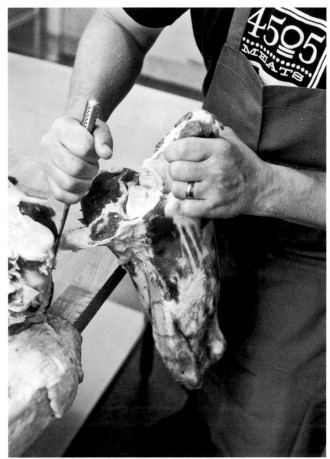

FOREQUARTER, SHANK, STEP 3: Putting pressure on the bottom of the shank will help release the socket from the joint.

FOREQUARTER, SHANK, STEP 4: While cutting, move the shank back and forth, and it will come right off.

FOREQUARTER, NECK, STEP 1: Now, flip over the forequarter. You can see the beautiful curvature in the spine where the neck begins and ends.

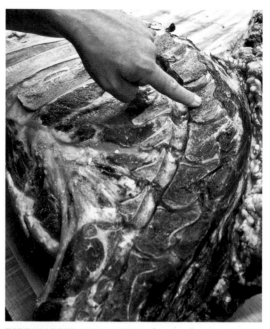

FOREQUARTER, NECK, STEP 2: Locate the planned cut. Saw through the 6th neck bone to separate the neck.

FOREQUARTER, NECK, STEP 3: Start your cut through the meat by resting your knife on the front of the arm bone.

FOREQUARTER, NECK, STEP 4: Continue cutting through all the meat to the other side before sawing.

FOREQUARTER, NECK, STEP 5: Saw through the bone to separate the neck from the chuck.

FOREQUARTER, NECK, STEP 6: Here, you can see that we sawed through the arm bone as well.

FOREQUARTER, NECK, STEP 7: A cross section of the chuck.

FOREQUARTER, NECK, STEP 8: **The finished bone-in neck section.**

FOREQUARTER, NECK MEDALLIONS, STEP 1: **Begin to de-bone the neck.**

FOREQUARTER, NECK MEDALLIONS, STEP 2: **Keep the knife as close to the bone as possible.**

FOREQUARTER, NECK MEDALLIONS, STEP 3: **Completely remove the neck bone.**

FOREQUARTER, NECK MEDALLIONS, STEP 4: **Remove the large cluster of fat.**

FOREQUARTER, NECK MEDALLIONS, STEP 5: **Remove the outside of the flap from the neck and begin to butterfly.**

FOREQUARTER, NECK MEDALLIONS, STEP 6: **Roll the neck evenly.**

FOREQUARTER, NECK MEDALLIONS, STEP 7: **Tie the neck. Make the first tie in the center and then continue to the ends, alternating ties from top to bottom, making the ties about 1½ inches/4 centimeters apart.**

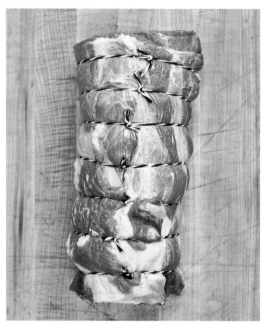

FOREQUARTER, NECK MEDALLIONS, STEP 8: The finished tied neck roast.

FOREQUARTER, NECK MEDALLIONS, STEP 9: Begin to portion. For even portions, make nice, clean, long cuts (avoid cutting with a sawing motion).

FOREQUARTER, NECK MEDALLIONS, STEP 10: Portion into 2-inch-/5-centimeter-thick cuts.

FOREQUARTER, NECK MEDALLIONS, STEP 11: The finished neck medallions. Wonderful when lightly braised (see page 82).

FOREQUARTER, RIB/CHUCK SECTION, STEP 1:
Measure the length of the loin on the forequarter, placing the tip of the knife where the loin starts and measuring to the edge of the meat.

FOREQUARTER, RIB/CHUCK SECTION, STEP 2:
From the bottom of the loin, add half of the length of the initial loin measurement to mark where you will begin to cut.

FOREQUARTER, RIB/CHUCK SECTION, STEP 3:
Make your mark on the rib side where the tip of the knife reached in the last step, which is 1½ times the length of the loin. Repeat the marking process on the other side.

FOREQUARTER, RIB/CHUCK SECTION, STEP 4:
Connect the two marks with a knife before beginning to saw.

FOREQUARTER, RIB/CHUCK SECTION, STEP 5: Move the chuck side around to prepare for sawing.

FOREQUARTER, RIB/CHUCK SECTION, STEP 6: Holding the rib firmly, saw through the ribs, following the knife markings as a guide.

FOREQUARTER, RIB/CHUCK SECTION, STEP 7: Pull one side up so that the saw can move freely; continue separating.

FOREQUARTER, RIB/CHUCK SECTION, STEP 8: Continue to cut through the blade bone, sawing only through the bone, and use your knife to cut through the flesh.

FOREQUARTER, PRIME RIB/UPPER CHUCK, STEP 1: The upper part of the forequarter. Amazing when roasted whole!

FOREQUARTER, PRIME RIB/UPPER CHUCK, STEP 2: Mark the point where you are going to separate the prime rib from the chuck, between the 5th and 6th ribs.

FOREQUARTER, PRIME RIB/UPPER CHUCK, STEP 3: Always plan your cut. Knowing your next move will result in a clean, perfect cut.

FOREQUARTER, PRIME RIB/UPPER CHUCK, STEP 4: **Begin the cut with the knife.**

FOREQUARTER, PRIME RIB/UPPER CHUCK, STEP 5: **Finish separating the rib from the chuck with the saw, sawing through the feather bones.**

FOREQUARTER, PRIME RIB/UPPER CHUCK, STEP 6: View from the feather-bone side, with the prime rib on the right and the upper chuck section on the left.

FOREQUARTER, PRIME RIB/UPPER CHUCK, STEP 7: Cross section of completed cuts: on the left is the prime rib; on the right is the upper chuck.

FOREQUARTER, UPPER CHUCK, STEP 1: Cubes of upper chuck, portioned for grinding.

FOREQUARTER, UPPER CHUCK, STEP 2: Grinding the chuck. Perfect for sausages (see pages 79-81) and burgers.

FOREQUARTER, PRIME RIB SECTION, STEP 1: **Now, working on the other prime rib, remove the feather bones and chine bones, as we did before on pages 26-27.**

FOREQUARTER, PRIME RIB SECTION, STEP 2: **Remove the plate meat from the bone-in prime rib.**

FOREQUARTER, PRIME RIB SECTION, STEP 3: Cut off the inedible gristle from the top of the rib.

FOREQUARTER, PRIME RIB SECTION, STEP 4: To make your beautiful bone-in rib eye chops, cut exactly half-way between each rib.

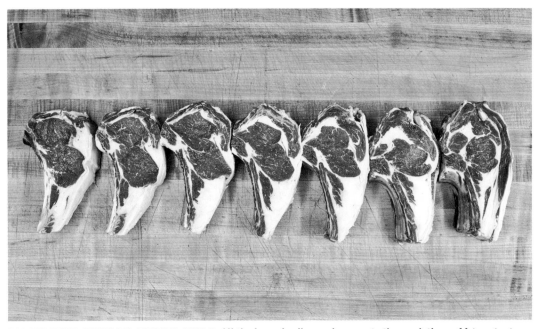

FOREQUARTER, PRIME RIB SECTION, STEP 5: All the bone-in rib eye chops; note the gradations of fat content. On the left is the short loin side of the chops; on the right is the chuck side. The closer we get to the short loin, the leaner the chops.

FOREQUARTER, LOWER CHUCK/PLATE SECTION, STEP 1: Count the ribs to locate the 5th and 6th ribs before making your cut.

FOREQUARTER, LOWER CHUCK/PLATE SECTION, STEP 2: Begin cutting through the plate section, exactly halfway between the 5th and 6th ribs.

FOREQUARTER, LOWER CHUCK/PLATE SECTION, STEP 3: Switch to the saw and continue to cut through the cartilage in the plate section.

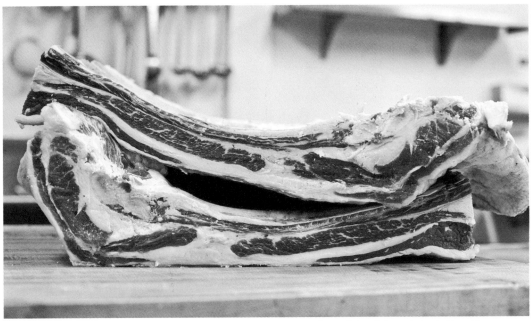

FOREQUARTER, LOWER CHUCK/PLATE SECTION, STEP 4: Cross sections of the plate and short rib sections.

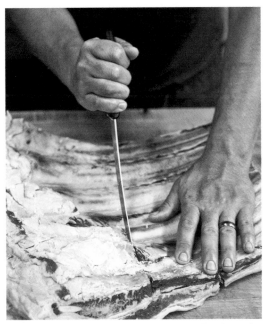

FOREQUARTER, PLATE SECTION, STEP 1: Cut through the flesh at the tip of the ribs, to separate the plate and the short rib sections.

FOREQUARTER, PLATE SECTION, STEP 2: Begin sawing in the same line where you cut the flesh.

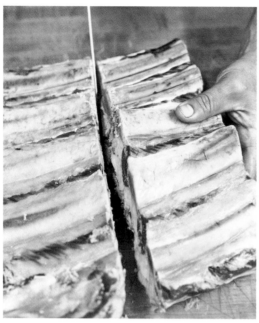

FOREQUARTER, PLATE SECTION, SHORT RIBS, STEP 1:
Finish dividing the short ribs with the knife.

FOREQUARTER, PLATE SECTION, SHORT RIBS, STEP 2:
Cut the rib meat in half so the ribs can easily fit in a pot.

FOREQUARTER, PLATE SECTION, SHORT RIBS, STEP 3: The finished short ribs on the bone. You'll have deeper flavor when you cook them on the bone; the bone is where the flavor lives.

FOREQUARTER, LOWER CHUCK SECTION, BRISKET, STEP 1: Begin cutting off the brisket from the top of the rib section, always keeping one side of the blade on the breastbone (this will result in a much higher yield).

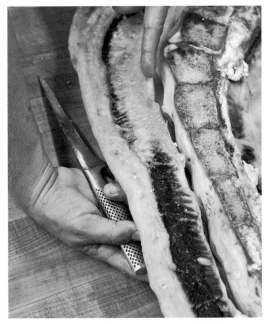

FOREQUARTER, LOWER CHUCK SECTION, BRISKET, STEP 2: Notice that the breastbone is beveled. Make your incision first and work your blade underneath the curvature of the breastbone.

FOREQUARTER, LOWER CHUCK SECTION, BRISKET, STEP 3: Now turn it around and come in from the other side to meet up with your first cut, and the ribs will begin to release.

FOREQUARTER, LOWER CHUCK SECTION, BRISKET, STEP 4: When the two cuts meet, the ribs will pull off easily, without cutting into the brisket.

FOREQUARTER, LOWER CHUCK SECTION, BRISKET,
STEP 5: Clean up a little bit of the fat from the brisket/
plate section.

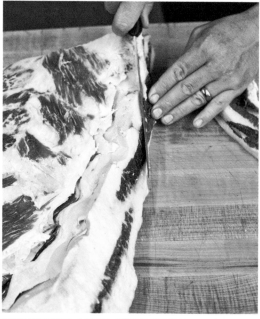

FOREQUARTER, LOWER CHUCK SECTION, BRISKET,
STEP 6: Trim off some of the oxidized (discolored and
dry) meat that was exposed to air while the meat was
hanging.

FOREQUARTER, LOWER CHUCK SECTION, BRISKET,
STEP 7: Divide the section in half lengthwise to separate the brisket from the plate section.

FOREQUARTER, LOWER CHUCK SECTION, BRISKET,
STEP 8: Continue to separate the brisket from the
plate.

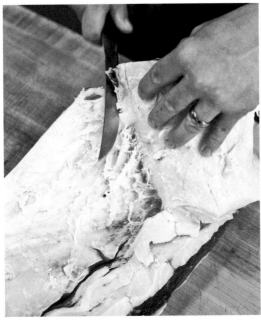

FOREQUARTER, LOWER CHUCK SECTION, BRISKET, STEP 9: Turn the brisket over and begin to pull off a bit more fat from the head (the front end of the brisket).

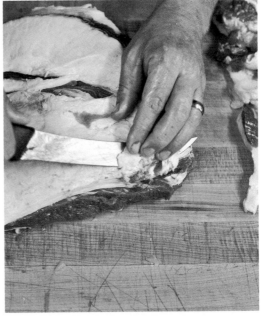

FOREQUARTER, LOWER CHUCK SECTION, BRISKET, STEP 10: Trim off some of the fat, but always leave on as much as possible. Fat is flavor; you can trim it off after the meat is cooked.

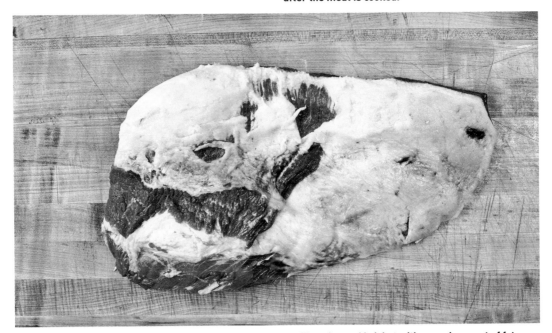

FOREQUARTER, LOWER CHUCK SECTION, BRISKET, STEP 11: The trimmed brisket with a good amount of fat.

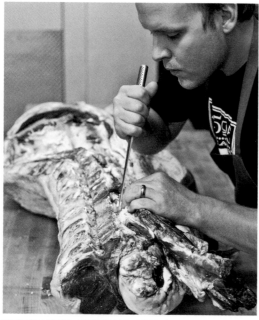

HINDQUARTER, SHORT LOIN SECTION, HANGER, STEP 1: Carefully remove the hanger steak from the short loin section.

HINDQUARTER, SHORT LOIN SECTION, HANGER, STEP 2: This is a trimmed and cleaned hanger steak, ready for cooking.

HINDQUARTER, FLANK SECTION, STEP 1: Release the first flap from the short loin section.

HINDQUARTER, FLANK SECTION, STEP 2: Pull off the big cluster of hard kidney fat. This will give you a better view so you can plan your next cut.

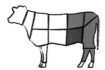

HINDQUARTER, FLANK SECTION, STEP 3: Now, remove the outside flap that we began to free up in step 1, cutting down the middle between the flank and the flap.

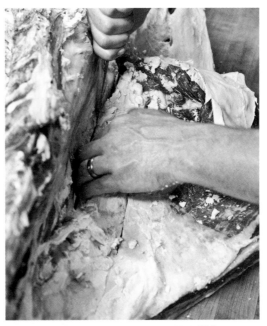

HINDQUARTER, FLANK SECTION, STEP 4: Make an incision along the tenderloin and the inside flap of the flank section.

HINDQUARTER, FLANK SECTION, STEP 5: Cut down alongside the 13th rib to remove the flap meat.

HINDQUARTER, FLANK SECTION, FLAP, STEP 1: Remove some of the fat from the flap meat. This is a very flavorful cut that's perfect for grilling.

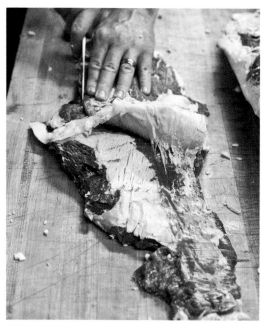

HINDQUARTER, FLANK SECTION, FLAP, STEP 2: Trim off some of the fat from the flap meat.

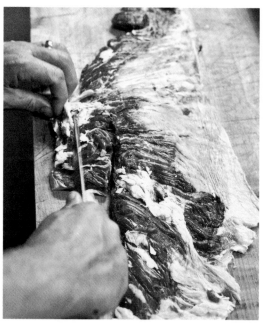

HINDQUARTER, FLANK SECTION, FLAP, STEP 3: Remove the sinew from the lower part; you can see all the beautiful grain in this cut. After cooking, be sure to slice against the grain.

HINDQUARTER, FLANK SECTION, FLANK STEAK, STEP 1: Cleaning up and pulling back the inedible sinew on the flank steak.

HINDQUARTER, FLANK SECTION, FLANK STEAK, STEP 2: Trim the fat from the flank steak.

HINDQUARTER, FLANK SECTION, FLANK STEAK, STEP 3: Trim off any remaining thick, inedible sinew.

HINDQUARTER, FLANK SECTION, FLANK STEAK, STEP 4: The finished flank steak.

HINDQUARTER, SHORT LOIN/SIRLOIN SECTION, STEP 1: Carefully pull back the butt end of the tenderloin from the sirloin, making small cuts and leaving behind as little meat as possible.

HINDQUARTER, SHORT LOIN/SIRLOIN SECTION, STEP 2: The butt end of the tenderloin, pulled back (this will protect the tenderloin when the short loin section is removed).

HINDQUARTER, SHORT LOIN/SIRLOIN SECTION, STEP 3: Cut all around the bone to avoid sawing through the meat. Doing this will preserve the option of leaving the tenderloin whole.

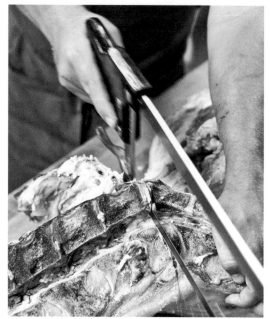

HINDQUARTER, SHORT LOIN/SIRLOIN SECTION, STEP 4: Follow your previous cut with the saw to free the short loin section from the sirloin.

HINDQUARTER, SHORT LOIN/SIRLOIN SECTION, STEP 5: **Finish the separation with the saw.**

HINDQUARTER, SHORT LOIN/SIRLOIN SECTION, STEP 6: **Come back and cut through the remaining meat with your knife.**

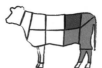

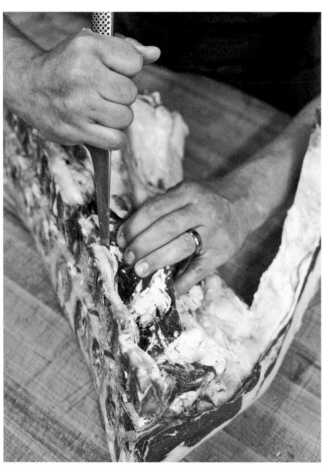

HINDQUARTER, SHORT LOIN SECTION, TENDERLOIN, STEP 1:
Remove the tenderloin from the short loin, starting along the chine bone and keeping one side of your knife touching the bone at all times.

HINDQUARTER, SHORT LOIN SECTION, TENDERLOIN, STEP 2:
Carefully, with almost a rolling motion, peel back the tenderloin.

HINDQUARTER, SHORT LOIN SECTION, TENDERLOIN, STEP 3:
Finish removing the tenderloin.

HINDQUARTER, SHORT LOIN SECTION, TENDERLOIN, STEP 4:
Remove a good amount of the cluster fat from the tenderloin, but
also leave on a generous amount of fat for great flavor. You can trim
it off before serving.

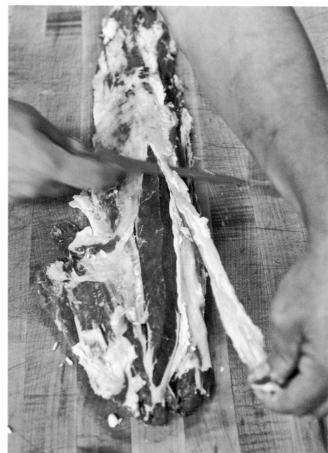

HINDQUARTER, SHORT LOIN SECTION, TENDERLOIN, STEP 5:
Remove any tough, thick, sinewy pieces from the tenderloin.

HINDQUARTER, SHORT LOIN SECTION, TENDERLOIN, STEP 6:
Remove more of the sinews and silver skin.

HINDQUARTER, SHORT LOIN SECTION, TENDERLOIN, STEP 7:
Tie the tenderloin every 2 inches/5 centimeters, starting in the center and then alternating ties from top to bottom to keep it even.

HINDQUARTER, SHORT LOIN SECTION, TENDERLOIN, STEP 8:
The whole tied tenderloin.

HINDQUARTER, SHORT LOIN SECTION, TENDERLOIN MEDALLIONS, STEP 1: Portion out the tenderloins into tenderloin medallions, also known as filets mignons.

HINDQUARTER, SHORT LOIN SECTION, TENDERLOIN MEDALLIONS, STEP 2: It's a good idea to keep some fat on lean cuts like this one. Fat can be removed later if you don't want to eat it.

HINDQUARTER, SHORT LOIN SECTION, TENDERLOIN MEDALLIONS, STEP 3: The finished tenderloin medallions.

HINDQUARTER, SHORT LOIN SECTION, NEW YORK STRIP, STEP 1: Remove the 13th rib: Begin by cutting around it, and then come back with the saw to finish the removal.

HINDQUARTER, SHORT LOIN SECTION, NEW YORK STRIP, STEP 2: With your knife, cut between the meat and bone to peel back the New York strip from the loin bones.

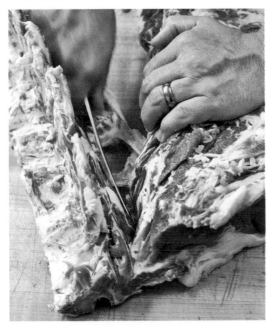

HINDQUARTER, SHORT LOIN SECTION, NEW YORK STRIP, STEP 3: Continue cutting all the way around the loin bones, keeping one side of the knife on the bone at all times.

HINDQUARTER, SHORT LOIN SECTION, NEW YORK STRIP, STEP 4: The New York strip released from the loin bones.

HINDQUARTER, SHORT LOIN SECTION, NEW YORK STRIP, STEP 5: Trim a little fat (but not much!) from the cap of the New York strip.

HINDQUARTER, SHORT LOIN SECTION, NEW YORK STRIP, STEP 6: Trim off the chine meat, which is really tasty but a little tough and chewy. This is great for tacos.

HINDQUARTER, SHORT LOIN SECTION, NEW YORK STRIP, STEP 7: The whole New York strip with a nice fat cap.

HINDQUARTER, SHORT LOIN SECTION, NEW YORK STRIP, STEP 8: Portion the New York strip into steaks, using nice, smooth strokes—no sawing back and forth. I always measure with two fingers, which makes about 2-inch-/5-centimeter-wide steaks.

HINDQUARTER, SHORT LOIN SECTION, NEW YORK STRIP, STEP 9: Portioned New York strip steaks; note the gradations of fat distribution. The right side is from the meat closest to the prime rib section; the left side is from the meat closest to the sirloin section.

HINDQUARTER, SHORT LOIN, PORTERHOUSE, STEP, 1: To cut the porterhouse, first remove the butt end of the tenderloin.

HINDQUARTER, SHORT LOIN, PORTERHOUSE, STEP 2:
The whole porterhouse roast.

HINDQUARTER, SHORT LOIN, PORTERHOUSE, STEP 3:
Mark the meat. (I always measure with two fingers, to cut thick, memorable 2-inch/5-centimeter-thick steaks.)

HINDQUARTER, SHORT LOIN, PORTERHOUSE, STEP 4:
First, cut through all the meat until the tip of the knife hits the bone.

HINDQUARTER, SHORT LOIN, PORTERHOUSE, STEP 5:
Then come back and saw through the bone to free up the steak.

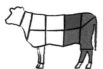

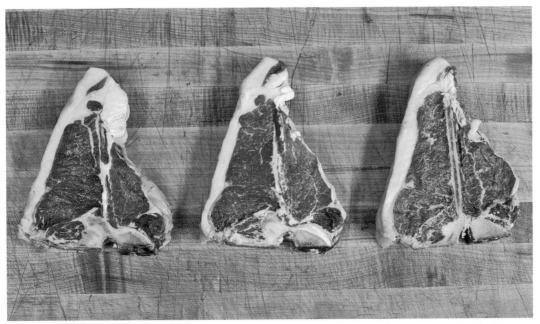

HINDQUARTER, SHORT LOIN, PORTERHOUSE, STEP 6: Porterhouse steaks have a large amount of tenderloin.

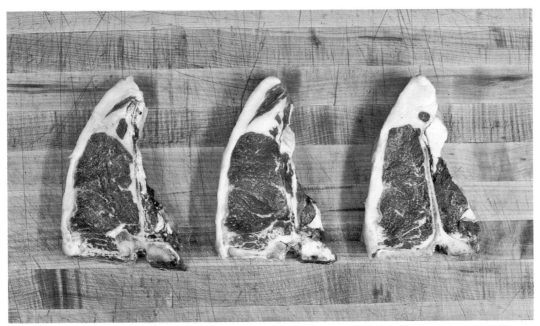

HINDQUARTER, SHORT LOIN, T-BONE: T-bone steaks are cut right next to the porterhouse and have a smaller amount of tenderloin than porterhouse.

HINDQUARTER, SIRLOIN/LEG SECTION, SPIDER,
STEP 1: **Begin to remove the spider cut on the inner side of the hip. It's perfect for a mid-butchering snack.**

HINDQUARTER, SIRLOIN/LEG SECTION, SPIDER,
STEP 2: **Cut on the other side of the spider cut that's underneath the fat.**

HINDQUARTER, SIRLOIN/LEG SECTION, SPIDER,
STEP 3: **With the knife flat on the bone, ease out the meat.**

HINDQUARTER, SIRLOIN/LEG SECTION, SPIDER,
STEP 4: **The spider cut is not a cut you'll see in the butcher's case. They are small, and there are only two per animal.**

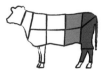

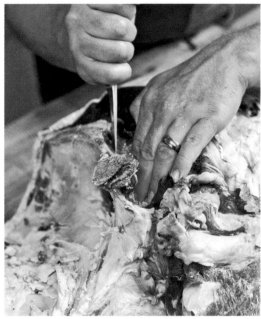

HINDQUARTER, SIRLOIN/LEG SECTION, STEP 1:
Remove the pelvic bone from the top of the hindquarter.

HINDQUARTER, SIRLOIN/LEG SECTION, STEP 2:
Cut along the aitch bone, pulling back on that bone and removing the meat from it, working the knife around the whole pelvic bone to the joint.

HINDQUARTER, SIRLOIN/LEG SECTION, STEP 3:
Cut through the ligament in the joint at the base of the pelvic bone, which is at the beginning of the leg bone.

HINDQUARTER, SIRLOIN/LEG SECTION, STEP 4:
Remove the whole pelvic bone, working it free.

HINDQUARTER, SIRLOIN/LEG SECTION, STEP 5:
Begin to remove the tip of the tri-tip from the top of the sirloin on the side of the leg.

HINDQUARTER, SIRLOIN/LEG SECTION, STEP 6:
Follow the crease to remove the tip of the tri-tip from the leg.

HINDQUARTER, SIRLOIN/LEG SECTION, STEP 7:
Start separating the sirloin from the round with your knife, making sure you don't cut through the tri-tip.

HINDQUARTER, SIRLOIN/LEG SECTION, STEP 8:
Cut the meat between the sirloin and the round right below the ball joint of the leg bone.

HINDQUARTER, SIRLOIN/LEG SECTION, STEP 9: Finish separating by sawing through the bone.

HINDQUARTER, SIRLOIN/LEG SECTION, TRI-TIP, STEP 1: The sirloin with tri-tip attached; remove a small cluster of fat to get a better view of the tri-tip.

HINDQUARTER, SIRLOIN/LEG SECTION, TRI-TIP, STEP 2: Separate the tri-tip from the bottom sirloin, at the natural seam of the muscles.

HINDQUARTER, SIRLOIN/LEG SECTION, TRI-TIP, STEP 3: Clean up the tri-tip, removing some but not all of the fat.

HINDQUARTER, SIRLOIN/LEG SECTION, TRI-TIP,
STEP 4: **Trim the edge of the tri-tip.**

HINDQUARTER, SIRLOIN/LEG SECTION, TRI-TIP,
STEP 5: **The finished tri-tip.**

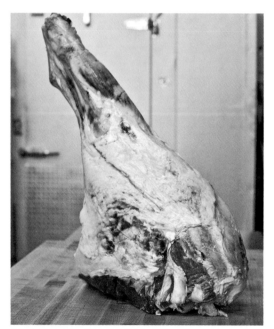

HINDQUARTER, LEG SECTION, STEAMSHIP, STEP 1:
The whole leg section.

HINDQUARTER, LEG SECTION, STEAMSHIP, STEP 2:
Cut through the tendon from the shank end.

HINDQUARTER, LEG SECTION, STEAMSHIP, STEP 3: Saw off the bottom part of the shin.

HINDQUARTER, LEG SECTION, STEAMSHIP, STEP 4: Make an incision at the top of the shank and continue all the way around, just below the bottom of the leg bone and shank joint.

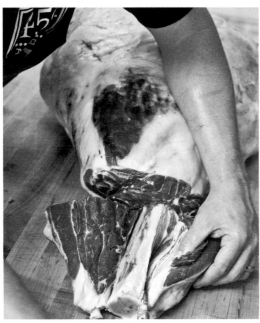

HINDQUARTER, LEG SECTION, STEAMSHIP, STEP 5: Remove the shank meat.

HINDQUARTER, LEG SECTION, STEAMSHIP, STEP 6: Scrape the bone with the back of your knife until it's nice and clean.

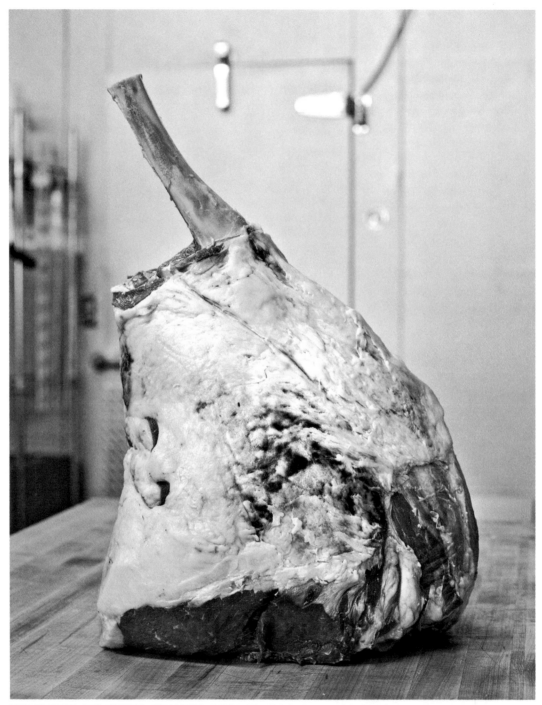

HINDQUARTER, LEG SECTION, STEAMSHIP, STEP 7: The finished steamship. This is not a common roast but one that is easy to execute and truly memorable. Just make sure you have a big enough oven or spit for spit roasting.

Red-Hot Beef Sausage

YIELD: 4 pounds/1.8 kilograms

These sausages are an ode to the famous Texas red hots. They are
really rich, with a beautiful, full flavor and a lot of heat. These sausages
should be hot-smoked until cooked through (to an internal temperature
of 148°F/64°C). If you're not equipped to hot-smoke, they are also
fantastic poached or gently grilled. To make the sausages, you
will need about 20 feet/6 meters of large or medium hog casings,
which can be ordered by the hank (about 100 feet/30 meters) from
a specialty butcher or on the Internet (see page 237).

Large or medium hog casings	About 20 feet/6 meters			
DRY INGREDIENTS				
Fine sea or kosher salt	1½ tbsp	1 oz	30 g	1.6%
Ground cayenne pepper	1½ tbsp	1 oz	30 g	1.6%
Sweet paprika	1 tbsp	0.2 oz	6 g	0.3%
Granulated sugar	⅓ cup	1.3 oz	38 g	2.1%
Mustard powder	1½ tbsp	0.6 oz	18 g	1%
Freshly ground black pepper	2 tsp	0.3 oz	7 g	0.4%
Ground coriander seeds	1 tsp	0.1 oz	3 g	0.2%
Chile powder	1½ tsp	0.2 oz	5 g	0.3%
MEAT				
Beef chuck and/or trim (see page 47), very cold	3.4 lb	55 oz	1,550 g	87.8%
WET INGREDIENTS				
Ice water	⅓ cup	3 oz	84 g	5.6%
Yellow mustard	1½ tbsp	0.3 oz	10 g	0.5%

cont'd

1. Read the General Sausage-Making Tips (see page 17).

2. The night before: Soak the hog casings in a bowl of cold water; refrigerate overnight.

3. Assemble all of the dry ingredients in a container. (This step need not be done the night before, but it's crucial that it be completed before you start grinding the meat.)

4. The next day: Untangle the casings and begin to open them to make the stuffing process easier. Hold one end of each piece of casing up to the nozzle of the faucet and support it with your other hand. Gently turn on the water and let it run through the casings to check for holes. If there are any holes in the casings, cut out the pieces with the holes. Hold the casings in a bowl of ice water or refrigerate until stuffing time.

5. With a sharp boning knife, or your knife of choice, remove the meat and fat from the bones, if necessary. Open-freeze the meat, uncovered, for 30 to 60 minutes, until the surface of the meat is crunchy to the touch and the interior is very cold, but not frozen.

6. Cut the beef into 1-inch-/2.5-centimeter-square cubes or a size slightly smaller than the opening of the meat grinder. Open-freeze the meat again, uncovered, for 30 to 60 minutes, until the surface of the meat is crunchy to the touch and the interior is very cold, but not frozen.

7. When you are ready to grind, prepare a perfectly clean and chilled meat grinder for grinding, and fit it with the medium plate. Start the auger and, without using the supplied pusher, let the auger gently grab each cube of meat and bring it forward toward the blade and through the grinding plate. Continue grinding until all of the meat has been processed. Place it in a clean, cold non-reactive bowl or tub and again open-freeze, uncovered, for 30 to 60 minutes, until the surface is crunchy to the touch and the interior is very cold, but not frozen.

8. In a medium nonreactive bowl, combine the dry ingredients with the ice water and yellow mustard and whisk together until completely blended and the dry ingredients have dissolved. I call this the "slurry."

9. In a large, wide basin or bowl that will give you plenty of room to mix the meat and seasonings, combine the cold meat with the slurry. Roll up your sleeves and, with perfectly clean hands, begin kneading and turning the mixture as you would a large quantity of bread dough. Eventually, you will begin to notice that the mixture has acquired a somewhat creamy texture. This is caused by the warmth of your hands and is a sign that you have finished mixing. Spoon out a few tablespoons of the mixture, and return the remainder to the refrigerator.

10. In a nonstick skillet over medium heat, lightly fry a test portion of sausage mixture until cooked through but not caramelized (which would change the flavor profile). Taste for seasoning. Based on this taste test, you can adjust the salt in the main portion of sausage, if desired.

11. Prepare a perfectly clean and chilled sausage stuffer and place the water-filled bowl of casings next to it. You will also need a landing surface of clean trays or parchment paper–lined baking sheets for your finished sausages.

12. Load the sausage mixture into the canister of the sausage stuffer, compacting it very lightly with a spatula to be sure there are no air pockets. Replace the lid.

13. Thread a length of casing all the way onto the stuffing horn and start cranking just enough to move a little of the ground meat mixture into the casing. As soon as you can see the meat poking through the nose of the stuffer, stop and crank backward slightly to halt the forward movement. Pinch the casing where the meat starts (to extrude all the air), and tie into a knot. Now start cranking again with one hand while you support the emerging sausage with the other. Move the casing out slowly to allow it to fill fully but not too tightly, so that there will be some give in the sausage when it comes time to tie the links. When you get close to the end, leave 6 inches/15 centimeters of unstuffed casing and stop cranking.

14. Go back to the original knot and measure 6 inches/15 centimeters of sausage. Pinch the sausage gently to form your first link, and twist forward for about seven rotations. Move another 6 inches/15 centimeters down the sausage, and this time, pinch firmly and twist backward. Repeat this process every 6 inches/15 centimeters, alternating forward and backward, until you reach the open end of the casing. Twist the open end right at the last bit of sausage to seal off the whole coil, and then tie a knot.

15. Ideally, hang the sausage overnight in a refrigerator, or refrigerate on parchment paper–lined baking sheets covered with plastic wrap, to allow the casing to form fully to the meat, and the sausage to settle. (Or, if desired, you can cook the sausages by smoking slow-and-low right away.) The next day, cut between each link and cook as desired.

Onion-Braised Beef Neck

SERVES 8

The neck is one of my favorite cuts from any animal. It's very flavorful yet still delicate, so you have to be careful not to over braise it. Adding white wine to this dish really takes it over the top, showcasing the flavor of the meat and at the same time cutting through the richness of the beef and cheese. I enjoy this braise over egg noodles; it's an especially tasty version of Beef Stroganoff. You can substitute any of the braising cuts, such as short ribs.

Beef neck (see page 38), about 2¼ in/5.5 cm thick	1	2.6 lb	1,189 g	21.4%
Fine sea salt (for initial seasoning)	1½ tsp	0.4 oz	10 g	0.2%
Coarsely ground pepper	as needed			
Grapeseed oil	as needed			
Yellow onion, cut lengthwise into 8 pieces	1 large	28.6 oz	810 g	14.5%
Carrots, peeled and cut into thick rounds	2 large	9.2 oz	260 g	4.6%
Fresh thyme sprigs, tied together	1 large bunch	1.6 oz	45 g	0.8%
Garlic, loose papery skin pulled off and top one-third trimmed off	1 whole head	1.6 oz	45 g	0.8%
Fine sea salt (for cooking)	1½ tsp	0.4 oz	10 g	0.2%
Dry white wine	1½ cups	12.7 oz	360 g	6.5%
Scallions, white and green parts, sliced and reserved separately	2 bunches	2.3 oz	65 g	1.2%
Unsalted chicken or vegetable stock (water works, too)	2½ qt	84.7 oz	2,401 g	43%
Whole black peppercorns	1 tsp	0.2 oz	7 g	0.1%

Dry egg noodles, cooked	½ lb	8 oz	226 g	4%
Cold unsalted butter, cut into pieces	½ cup	4 oz	112 g	2%
Freshly grated Parmesan	6 tbsp	1.4 oz	40 g	0.7%

1. Preheat the oven to 325°F/165°C/gas 3. Butterfly the neck to form a rectangle (see page 37). Season all over with the salt and lightly with pepper. Roll into a cylinder and make the first tie in the center. Continue, alternating ties above and below to form a firm cylinder about 4 inches/10 centimeters in diameter. Cut between the ties into approximately 3½-inch-/9-centimeter-thick medallions.

2. In a large ovenproof braising pot or Dutch oven, heat 1 to 2 tablespoons of grapeseed oil (depending on the size of the pot) over medium-high heat until very hot but not smoking. Add the medallions and sear until nicely colored on all sides.

3. Scatter the onion, carrots, and thyme over the meat. Halve the head of garlic lengthwise and place the halves, cut-side down, directly on the hot surface of the pot. Cook for 5 minutes, and then season with the salt.

4. Add the wine and bring to a boil. Add the whites from the scallions, the stock, and the peppercorns, and bring to a simmer. Cover the pot with a large sheet of parchment paper, pressing it down so that it touches the surface of the liquid and extends up the sides and over the rim of the pot. Place the lid over the parchment and transfer the pot to the oven. Braise for about 2½ hours, until the meat is very tender. (Check every 30 to 45 minutes and lower the heat by a few degrees if the liquid is simmering briskly—it should barely quiver.)

5. Remove the pot from the oven. Lift off the parchment and discard it. Remove all the meat and the garlic, setting them aside in a bowl. Strain the braising juices into a large saucepan and discard the remaining solids. Squeeze the softened garlic cloves into the braising juices and simmer actively over medium-high heat, until reduced by about half.

6. Meanwhile, pull apart the meat into bite-size chunks and add them to the pan with the reduced juices. Simmer for 5 minutes, until warm.

7. Add the noodles, scallion greens, butter, and half of the Parmesan to the pan; stir gently. Remove the pan from the heat. Taste and adjust the seasoning.

8. Ladle the mixture into bowls, distributing all the tasty morsels equally. Sprinkle with the remaining Parmesan.

Ancho and Kaffir
Beef Flintstone Chop

· ·

SERVES 4

This massive chop is made up of the bone-in rib eye with the short rib still attached. Some people french the rib (the result looks a little like a tennis racket), but I like to leave the rich rib meat attached. Normally, short ribs are braised rather than served on the pink side, as I do here, but I love the delicious, chewy rib meat along with the perfectly medium-rare, beefy rib eye. This home oven–friendly technique is very simple, and yields an exterior that is charred and crisp with an interior that remains a uniform juicy pink from top to bottom. Use this technique with any favorite steak—it's all about temperature control in the oven and using the thermometer rather than the timer to decide when it's done.

ANCHO RUB				
Ancho chiles, stemmed	2 whole	0.9 oz	25 g	1.4%
Dried chipotle chiles, stemmed	½ whole	0.2 oz	5 g	0.3%
Whole black peppercorns	½ tsp	0.1 oz	4 g	0.2%
Kaffir lime leaves	2 to 3	0.1 oz	2 g	0.1%
Coriander seeds	1 tsp	0.2 oz	5 g	0.3%
Fine sea salt	2¼ tsp	0.5 oz	14 g	0.7%
MEAT				
Flintstone chop (see page 31) 2½ in/6 cm thick and 18 in/46 cm long, with the rib meat still attached	1 whole	64.4 oz	1,825 g	97%

1. In a spice grinder, combine the chiles, pepper-corns, lime leaves, and coriander seeds and pulverize them into a fine dust. Blend in the salt. Rub and massage the spice mixture all over the chop, including the fatty edge and the rib meat, patting it to help it stick. Bring the chop to room temperature, about 1 hour.

2. Preheat the oven to 250°F/120°C/gas 1. Place a rack over a roasting pan (or roast directly on the oven rack with a drip pan underneath) and place the chop on the rack. Insert a probe thermometer into the eye of the meat about ½ inch/12 millimeters away from the bone; place the controls outside the oven. Slow-roast the chop for 1 to 1½ hours, or until the internal temperature reaches 132°F/55°C for medium-rare or 142°F/61°C for medium, whichever you prefer. Set the alarm on your probe thermometer, if it has this feature.

3. When the temperature reaches about 100°F/38°C, start a hardwood or hardwood charcoal fire and let it burn down for medium-high–heat grilling, or preheat a gas grill.

4. Grill the chop briefly, just to add grill marks and a nice, charred flavor. When the lovely chile-tinged fat on the cap begins to crack, render, and fry itself, the chop is ready. (Remember that the inside is already cooked perfectly.) Let the chop rest for 5 to 10 minutes, then cut it into thin slices and enjoy.

Spice-Cured Beef Brisket

YIELD: about 7 pounds/3 kilograms

For this recipe, you will make up the entire amount of the curing
mixture, then divide it in half and use the first half to cure the brisket
for 4 days, after which the cure becomes damp and somewhat crusty
because of the moisture that is leached out of the meat. To be fully cured,
however, the brisket will need 4 days more. So, discard the first batch
of cure and replace it with the remaining half of nice, dry cure.
This brisket is an amazing substitute for *bresaola* (which is made from
the leg); in fact, I like it better. Because of the marbling and the
grain, it falls apart in your mouth and has a richer flavor and
mouthfeel because of that awesome fat.

DRY CURE				
Coriander seeds	1½ cups	4.4 oz	126 g	1.3%
Caraway seeds	2½ cups	8.9 oz	252 g	2.5%
Crushed chile flakes	1½ cups	4.4 oz	126 g	1.3%
Whole black peppercorns	⅔ cup	3.5 oz	100 g	1%
Fine sea salt	6 lb	96 oz	2,722 g	27.6%
Garlic powder or granulated garlic	1⅔ cups	7.9 oz	224 g	2.3%
Dextrose or granulated sugar	3 lb	48 oz	1,361 g	13.8%
Curing salt (also known as pink salt, or sodium nitrate)	⅔ cup	6.5 oz	184 g	1.9%
Onion powder or granulated onion	2 cups	7.9 oz	224 g	2.3%
MEAT				
Beef brisket (see page 55), untrimmed	10 lb	10 lb	4,537 g	46%

1. In a spice grinder, combine the coriander seeds, caraway seeds, chile flakes, and peppercorns and process until coarsely ground. In a large mixing bowl, combine the salt, garlic powder, dextrose, curing salt, onion powder, and ground spice mixture and stir until thoroughly blended. Divide the cure in half. Transfer half of the cure to a tightly sealed container and set it aside in a cool, dry place. (This is the cure you will use for the second 4-day curing period.)

2. Pour half of the remaining cure into a deep nonreactive container that will fit in your refrigerator and put the brisket on top. Pour the remaining cure over the brisket and massage it deeply into both sides of the meat, rubbing it in well. Completely cover the container with several overlapping layers of plastic wrap, running it around the base of the container several times to be sure it is airtight.

3. Cure the brisket in the refrigerator without disturbing it for 2 days, then flip the brisket over and rewrap. Again, cure without disturbing for 2 days, for a total of 4 days.

4. After 4 days, the salt cure will be damp and crusty; scrape off the salt clusters, loosening the cure, and discard. Clean and dry the container, then repeat with the reserved curing mixture exactly as before, for a total of 8 days curing time.

5. Remove the brisket from the container and brush off all the spice mix and salt clusters. Hang in a sanitized and bacteria-free environment at 60° to 70°F/15°C to 21°C with a relative humidity of 65 percent to 75 percent for 90 days, or until the brisket is completely firm.

6. Trim off any dry large clusters of salt and/or pieces that appear inedible. Slice very thinly against the grain, and enjoy. (To store, wrap in butcher paper and store in the refrigerator for up to two weeks or Cryovac and seal airtight and store in the refrigerator for up to 2 months. Note: If there is any gray, black, green, or other color of mold, or an "off" smell, discard the whole piece. It's not worth the risk.)

Beef Tongue Pastrami

· ·

YIELD: about 3 pounds/1.36 kilograms

Since pepper is by nature coarse and insoluble, it does not penetrate
meat as easily as salt and sugar. Therefore, you've got to be really
aggressive with the flavor here to get through that thick skin.
When it comes to simmering the tongue, there is so much flavor
already that I use water, but you can definitely use beef stock—hey,
it would just liven up the party. I love to grill the tongue and serve
it hot in big pieces or cold in thin slices.

DRY CURE				
Fine sea salt	3 lb	48.1 oz	1,362 g	22.1%
Dextrose or granulated sugar	3 cups	24 oz	681 g	11.1%
Black peppercorns, coarsely ground	⅔ cup	3.5 oz	100 g	1.6%
MEAT				
Grass-fed beef tongue, rinsed and patted dry	3½ lbs	56.3 oz	1,594 g	25.9%
BRINE				
Master Brine (see page 16), completely cold	10 cups	80.1 oz	2,270 g	36.9%
Black peppercorns, finely ground	1 cup	5.2 oz	150 g	2.4%

1. In a spice grinder, combine the salt, dextrose, and coarsely ground peppercorns and process until finely ground.

2. Put half the cure into a nonreactive container (only just large enough to fit the tongue), and place the tongue on top. Dump the remaining cure on top and massage the mix deeply into both sides of the meat, rubbing it in well. Cover the container thoroughly with several layers of overlapping plastic wrap, running it around the base of the container several times to be sure it is airtight.

3. Cure the tongue in the refrigerator without disturbing it for 2 days, then flip the tongue over, rub once more with the cure, and rewrap as before. Again, cure without disturbing, for 2 days longer.

4. After 4 days, the salt cure will be slightly wet and crusty; scrape off the salt clusters, loosening the cure, and discard.

5. Combine the Master Brine and finely ground peppercorns in a tall, narrow nonreactive container only just large enough to hold the tongue upright and submerged. Once the brine is completely cold, add the tongue and lay a sheet of parchment paper over the top so that no part of the meat will be exposed to oxygen; weight with a sterilized plate to keep the meat submerged. Cure in the refrigerator for 8 days.

6. Pat the tongue dry. If you like, hot-smoke the tongue over hickory chips at 275°F/135°C/gas 1 for 4 hours before the next step.

7. Place the tongue in a large pot and cover with water. Simmer very gently for 6 hours, topping off the water when necessary to keep the tongue submerged. When it has finished cooking, the tongue should be nice and tender. Leave it in the cooking liquid, and let cool to room temperature. Transfer the tongue, still in the liquid, to the refrigerator and let stand overnight (this resting time helps the smoky flavor fully penetrate the meat).

8. Discard the water, which will have a very strong salty-smoky flavor (or use it as *part* of the liquid for a tongue soup). To serve, peel off the skin and cut it crosswise into thin slices. Enjoy!

CHAPTER TWO

LAMB

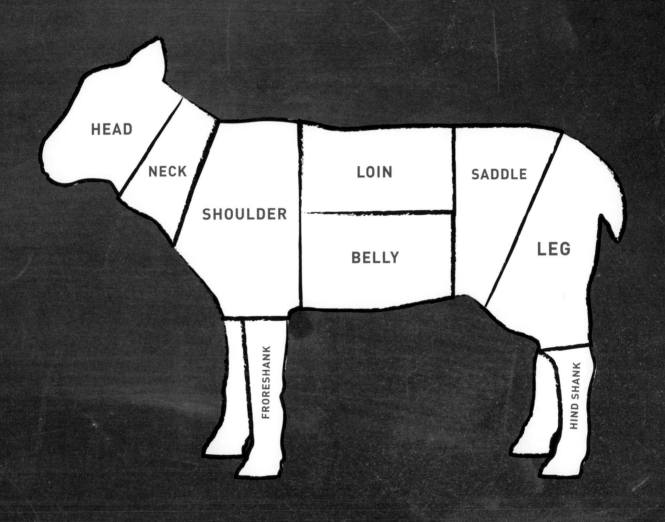

Breaking down a lamb is

much easier than butchering a steer or a pig, simply because it's a smaller animal and easier to maneuver. Although the anatomy is the same, we usually break down a lamb slightly differently. For one thing, each cut is much smaller—think about the difference in the sizes of a beef tenderloin, a pork tenderloin, and a lamb tenderloin. This animal was young at the time of slaughter, so the meat isn't very fatty and care must be taken when cooking to add fat—whether it's fat from other parts of the lamb or from butter or cured pork, like bacon or lardo.

The **neck** is full of fat and flavor, perfect for merguez (sausage), and the **shanks** are an easy choice for braising. The shoulder runs from the 1st to the 4th rib. On one side, I've portioned bone-in shoulder chops. On the other side, I've removed the shank and neck to create a bone-in roast. Any pieces of meat I have left over will go into the merguez. All of the meat from this whole area takes well to light braising and slow roasting.

I love cooking **shoulder chops** on a grill over slow-and-low heat. They're not the most tender, but they're not really tough. Get a good char on the fat for flavor. If you're preparing a **bone-in shoulder roast,** it's better to slow-roast it (at 225°F/100°C) for a good amount of time, until it reaches an internal temperature of 180°F/82°C and the meat is falling off the bone. Or you can debone it first, tie it up, marinate it, and then sear the outside before you switch to slow-and-low heat. Welcome to Flavor Town.

The loin and belly section goes from the 5th to the 13th rib. On one side, I've created a **frenched rack,** and with the extra belly you can make lamb pancetta—super tasty. There is not a lot of it, but it's one of the best cuts on the animal. It's a little chewy, but it has a lot of flavor.

Here's a key tip: Keep the **chine bone** on if you plan to roast the rack. The tip of the rib and the chine bone should both rest on the roasting pan, keeping the delicate eye of the loin safely away from the hot pan.

My **"lambchetta"** is truly awesome. I remove all the bones from the loin and wrap the loin in the belly (which protects the lean loin meat), to create a small porchetta-type roast. You can grill it whole until crispy, and then switch to slow-and-low heat and roast until 132°F/55°C. Using this technique, the belly acts as an impenetrable barrier, trapping the juices so they can't escape through that belly. You'll end up with crunch and flavor on the outside, juiciness and tender love on the inside. Or portion it out into medallions, then I sear in a pan with rosemary and garlic.

When cooking a lean cut of lamb, it's important to roast it to an internal temperature of 132F°/55°C. This is the true key to making really tasty lamb.

Once I've reached the **saddle section,** I make medallions from one side of the animal, and beautiful bone-in porterhouse lamb steaks from the other side. The medallions are cut from the **boneless loin** cut.

Don't try to remove every bit of fat from the **tenderloin** or the boneless loin. There is a lot of lean meat here, and it's important to cook it slow-and-low; this will help keep this meat from drying out. I eat lamb rare to medium-rare if it's not braised. With the **porterhouse**, I like to put it on the grill, brushing it with melted herb butter or rendered lamb fat with a branch of rosemary. The **sirloin** will have to cook slow-and-low after the first sear, even if it ends up on the grill. Baste with garlic and thyme and, of course, fat, fat, fat.

In the pictures here, I proceed straight to tying up the roasts after they've been boned out, but if they were really going to be cooked, I'd add flavor with brine or rub, and so should you.

Meat from the **leg** is pretty lean, but it has a lot of good flavor. It's smaller than a leg of pork, so it's easier to manage. My favorite is slow-and-low in a wood oven for a couple of hours, covered in herbs and basted with fat, which brings out a lot of flavor.

With the other leg, I again remove the bones, but this time I break it down following the natural separations of the meat, dividing them into five to seven individual muscles, which I love to stuff and wrap in pork belly, bacon, or caul fat. (These muscles are all lean, so it's all about figuring out how *not* to overcook them; see my recipe on page 142 for one way.) When you are cutting the leg into different, smaller, individual muscles, concentrate on only one at a time.

The whole leg is good for a crowd or for the family, while these smaller muscles are good for just a few diners. When grilling, baste it with the herb butter or rendered lamb fat. I love the flavor of **lamb fat** when it has been roasted nicely. I always sear the fat side first to develop its flavor, *then* let it render down slowly.

Obviously, I am a huge fan of fat and all the flavor it brings to the table. Some of my favorite preparations are served on fresh veggies and greens. Even though I love cooking with, in, and on fat, I always balance it with a fresh and bright note to cut the richness.

WHOLE ANIMAL: Our beautiful whole 58-pound/ 26-kilogram pasture-raised lamb.

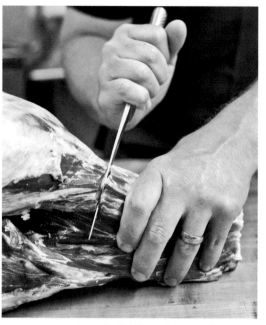

NECK SECTION, STEP 1: Cut through the flesh around the neck bone with the knife.

NECK SECTION, STEP 2: Come back with the cleaver and chop through the neck bone.

NECK SECTION, STEP 3: The bone-in lamb neck.

SHOULDER/LOIN SECTION, STEP 1: **Make an incision between the 5th and 6th ribs to remove the shoulder section.**

SHOULDER/LOIN SECTION, STEP 2: **Continue the cut upward between the two ribs, exactly halfway between the bones.**

SHOULDER/LOIN SECTION, STEP 3: **Repeat the process on the other side with the knife, cutting through the flesh only.**

SHOULDER/LOIN SECTION, STEP 4: **Now, with the saw, complete the separation by sawing through the backbone.**

SHOULDER SECTION, STEP 1: The fully removed shoulder section.

SHOULDER SECTION, STEP 2: Split apart the shoulder section with a saw, cutting right down the center of the backbone.

SHOULDER SECTION, STEP 3: Never saw flesh; stop sawing as soon as you go completely through the bone.

SHOULDER SECTION, STEP 4: Come back with the knife to cut through the remaining flesh.

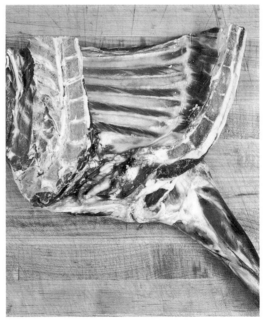

SHOULDER SECTION, STEP 5: The shoulder, with the foreshank and breast on.

SHOULDER SECTION, STEP 6: Cut through the cartilage on the breast at the tip of the ribs.

SHOULDER SECTION, STEP 7: Fully remove the breastbone.

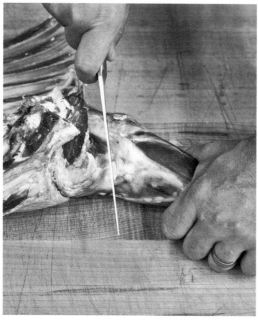

SHOULDER SECTION, SHANK, STEP 1: Cut all the way around the arm bone with the knife.

SHOULDER SECTION, SHANK, STEP 2: Come back with the saw and cut through the bone to remove the foreshank.

SHOULDER SECTION, SHOULDER CHOPS, STEP 1: With the saw, mark between each rib.

SHOULDER SECTION, SHOULDER CHOPS, STEP 2: Saw only through the backbone between each rib.

SHOULDER SECTION, SHOULDER CHOPS, STEP 3: Come back with the knife to cut through the remaining flesh.

SHOULDER SECTION, SHOULDER CHOPS, STEP 4: The finished bone-in shoulder chops.

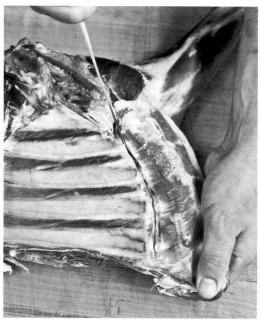

SHOULDER SECTION, BONE-IN SHOULDER, STEP 1:
Again, remove the breastbone, but now from the other shoulder. Cut through the cartilage at the tip of the ribs, then fully remove the breastbone.

SHOULDER SECTION, BONE-IN SHOULDER, STEP 2:
Cut through with the knife, then saw to remove the foreshank.

SHOULDER SECTION, BONE-IN SHOULDER, STEP 3:
The squared-off bone-in shoulder.

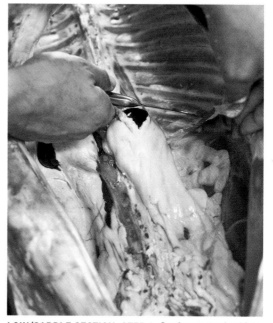

LOIN/SADDLE SECTION, STEP 1: Cut between the 12th and 13th ribs to remove the loin and belly section from the rest of the lamb.

LOIN/SADDLE SECTION, STEP 2: **Repeat with the knife on the other side of the rib cage.**

LOIN/SADDLE SECTION, STEP 3: **As you cut, be sure to follow the angle of the ribs.**

LOIN/SADDLE SECTION, STEP 4: **Turn the lamb on its side and saw through the backbone to remove the belly and loin section.**

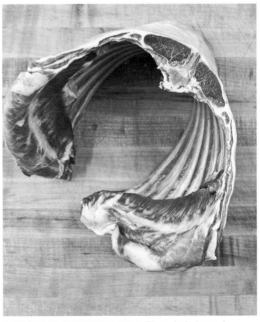

LOIN/BELLY SECTION, STEP 1: **The whole bone-in loin and belly section.**

LOIN/BELLY SECTION, STEP 2: Saw down through the backbone only, exactly in the center.

LOIN/BELLY SECTION, STEP 3: Then come back with the knife and finish cutting through the flesh.

LOIN/BELLY SECTION, STEP 4: One side of the bone-in loin section.

LOIN/BELLY SECTION, FRENCHED LAMB RACK, STEP 1: Remove the inside flap (on the steer, this would be the skirt steak).

LOIN/BELLY SECTION, FRENCHED LAMB RACK, STEP 2:
At the tip of the rack, cut through the soft cartilage and fully remove the cartilage section.

LOIN/BELLY SECTION, FRENCHED LAMB RACK, STEP 3:
Cut directly into the center back of each rib bone, cutting through the sinew only.

LOIN/BELLY SECTION, FRENCHED LAMB RACK, STEP 4:
Peel off the sinew from the tip of the rib.

LOIN/BELLY SECTION, FRENCHED LAMB RACK, STEP 5:
It's very important to peel the sinew about ½ in/ 5 centimeters down from the tip of each rib before you move on to the next rib.

LOIN/BELLY SECTION, FRENCHED LAMB RACK, STEP 6:
Make a "C" with your index finger, and pull down toward the eye of the loin to remove *all* the sinew.

LOIN/BELLY SECTION, FRENCHED LAMB RACK, STEP 7:
The frenched bones protrude from the rib meat.

LOIN/BELLY SECTION, FRENCHED LAMB RACK, STEP 8:
Come back with the knife to cut through to remove all the rib/belly meat.

LOIN/BELLY SECTION, FRENCHED LAMB RACK, STEP 9:
The whole frenched rack with the chine on. (It's important to keep the chine on if you plan to roast the rack. The tip of the rib and the backbone should both rest on the roasting pan, keeping the delicate eye of the loin safely away from the direct heat of the hot pan.)

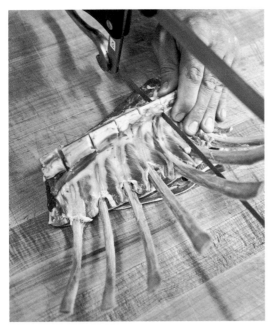

LOIN/BELLY SECTION, FRENCHED RIB CHOPS, STEP 1:
Saw through the top of the chine bone between each rib.

LOIN/BELLY SECTION, FRENCHED RIB CHOPS, STEP 2:
Come back with the knife to cut through and release each chop.

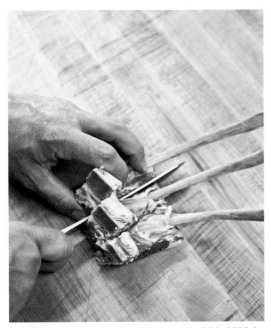

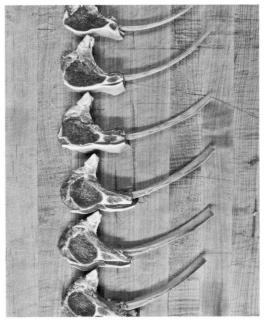

LOIN/BELLY SECTION, FRENCHED RIB CHOPS, STEP 3:
It's very important to secure the rack with one hand so that you can make a nice clean cut. Avoid a back-and-forth sawing motion, which will damage the delicate meat.

LOIN/BELLY SECTION, FRENCHED RIB CHOPS, STEP 4:
The finished frenched lamb rib chops.

LOIN/BELLY SECTION, LAMBCHETTA, STEP 1: On the other side of the belly, remove the inside flap (see page 102) as before. Cut through the soft cartilage at the tip of the ribs without going through the belly.

LOIN/BELLY SECTION, LAMBCHETTA, STEP 2: Make a shallow cut on each side of each rib, again without puncturing the belly.

LOIN/BELLY SECTION, LAMBCHETTA, STEP 3: Come back underneath the rib to remove the rib bone and sinew, leaving behind as much meat as possible.

LOIN/BELLY SECTION, LAMBCHETTA, STEP 4: Continue to release the rib bones, pulling up and keeping the knife as close to the bone as possible.

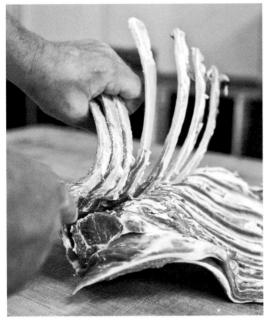

LOIN/BELLY SECTION, LAMBCHETTA, STEP 5: Continue to pull and carefully cut everything off the bones, cutting down toward the backbone.

LOIN/BELLY SECTION, LAMBCHETTA, STEP 6: Finally, remove all of the bones, keeping the side of the knife on the bone at all times to maximize your yield of loin meat.

LOIN/BELLY SECTION, LAMBCHETTA, STEP 7: Once the rib bones have been removed, come back and remove the tip of the shoulder blade (soft cartilage) that is still on the belly.

LOIN/BELLY SECTION, LAMBCHETTA, STEP 8: The finished boneless belly and loin section.

LOIN/BELLY SECTION, LAMBCHETTA, STEP 9: **Roll up tightly, with the loin on the inside.**

LOIN/BELLY SECTION, LAMBCHETTA, STEP 10: **Make the first tie on the lambchetta roast in the center. Make the second tie below and the third tie above the center tie, to keep the diameter of the roast even.**

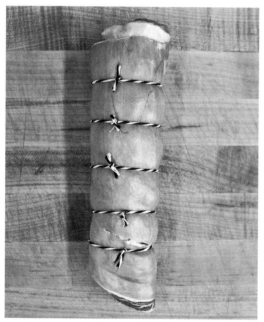

LOIN/BELLY SECTION, LAMBCHETTA, STEP 11: **The finished, tied lambchetta—grilling this cut whole is always my first choice!**

LOIN/BELLY SECTION, LAMBCHETTA, STEP 12: **Begin portioning out the lambchetta medallions, cutting between the strings. Make smooth, even cuts; do not saw back and forth with the knife.**

LOIN/BELLY SECTION, LAMBCHETTA, STEP 13: **The finished lambchetta medallions, 5 to 6 ounces/142 to 170 grams each.**

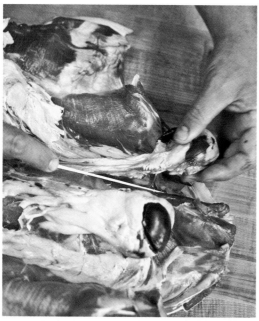

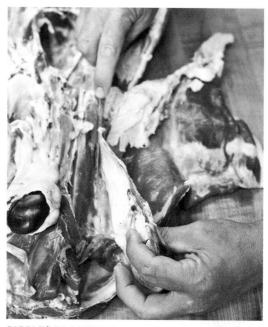

SADDLE/LEG SECTIONS, KIDNEY, STEP 1: Remove the kidney from the saddle section, cutting carefully through the surrounding kidney fat to avoid cutting through the tenderloin.

SADDLE/LEG SECTIONS, KIDNEY, STEP 2: Pull back the kidney fat with the kidneys inside, making just a few cuts.

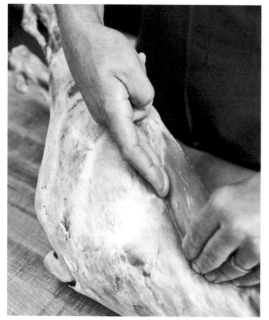

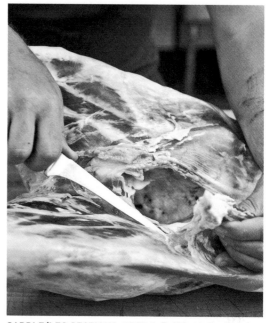

SADDLE/LEG SECTIONS, STEP 1: Here is the crease between the leg and the saddle flap, where we will separate the saddle and leg sections.

SADDLE/LEG SECTIONS, STEP 2: Pull back the inner part of the saddle flap, which is attached to the leg section in two places. Cut through these two areas to remove the flap.

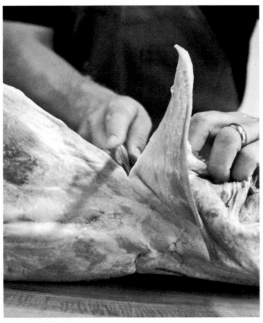

SADDLE/LEG SECTIONS, STEP 3: Remove the outer saddle flap from the leg section on both sides.

SADDLE/LEG SECTIONS, STEP 4: Cut all the way down to the pelvic bone.

SADDLE/LEG SECTIONS, STEP 5: Repeat this step on the other side of the saddle.

SADDLE/LEG SECTIONS, STEP 6: Cut all the way down through the butt of the tenderloin and flap, stopping at the pelvic bone.

SADDLE/LEG SECTIONS, STEP 7: **Come back with the saw to cut through the pelvic bone.**

SADDLE/LEG SECTIONS, STEP 8: **Use the knife to finish cutting through the remaining flesh.**

SADDLE SECTION, STEP 1: **Trim away the fat and any sinewy pieces left over from the kidneys.**

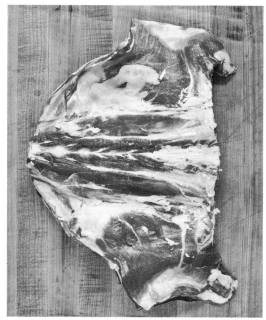

SADDLE SECTION, STEP 2: **The finished bone-in saddle.**

SADDLE SECTION, STEP 3: Saw through the center of the backbone to split the saddle.

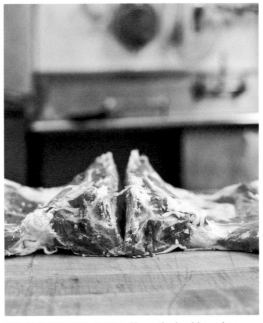

SADDLE SECTION, STEP 4: Here, the backbone has been sawed through, nice and even.

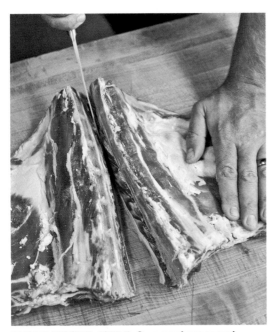

SADDLE SECTION, STEP 5: Once you have sawed through the bone, come back with the knife and finish cutting through the flesh to separate the two halves of the saddle.

SADDLE SECTION, STEP 6: The two split, bone-in saddle sections.

SADDLE SECTION, PORTERHOUSE, STEP 1: **Remove the 13th rib, also known as the floating rib.**

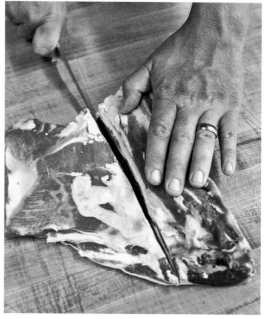

SADDLE SECTION, PORTERHOUSE, STEP 2: **Remove the saddle flap to square up the porterhouse section.**

SADDLE SECTION, PORTERHOUSE, STEP 3: **Cut through the flesh only to portion the porterhouse as desired.**

SADDLE SECTION, PORTERHOUSE, STEP 4: **Now we're ready to chop through the backbone.**

SADDLE SECTION, PORTERHOUSE, STEP 5: **Come back with the cleaver and a mallet; position the cleaver blade straight down on the backbone with the mallet resting on top.**

SADDLE SECTION, PORTERHOUSE, STEP 6: **Without moving the cleaver, carefully tap—using some force— with the mallet to cut through the backbone.**

SADDLE SECTION, PORTERHOUSE, STEP 7: **(left to right) The finished, bone-in lamb porterhouse and T-bone steaks.**

SADDLE SECTION, TENDERLOIN, STEP 1: Begin carefully removing the tenderloin from the saddle section.

SADDLE SECTION, TENDERLOIN, STEP 2: Pull back the tenderloin and make shallow cuts to remove it, leaving as much meat as possible on the tenderloin.

SADDLE SECTION, TENDERLOIN, STEP 3: The finished lamb tenderloin with a good amount of the tasty fat left on.

SADDLE SECTION, BONELESS LOIN, STEP 1: **Come back and start removing the backbone, cutting underneath the bone.**

SADDLE SECTION, BONELESS LOIN, STEP 2: **Continue removing the bone, leaving as much meat as possible on the loin.**

SADDLE SECTION, BONELESS LOIN, STEP 3: **Keeping one side of the knife always on the bone, finish removing the backbone.**

SADDLE SECTION, BONELESS LOIN, STEP 4: **The finished, whole lamb strip loin.**

SADDLE SECTION, BONELESS LOIN, STEP 5: Begin to portion out the strip loin with nice smooth cuts.

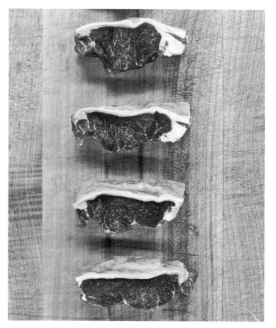

SADDLE SECTION, BONELESS LOIN, STEP 6: The finished, boneless lamb strip steaks.

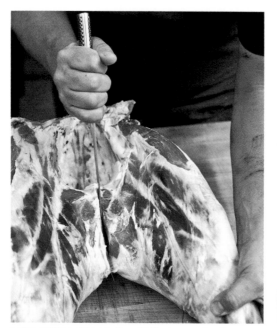

LEG SECTION, STEP 1: To split the leg section into halves, cut through the flesh with your knife first.

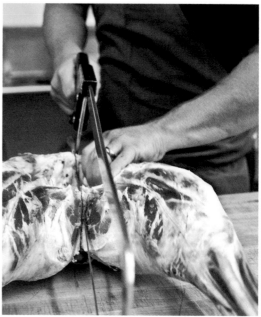

LEG SECTION, STEP 2: Come back with a saw to cut through the center of the pelvic bone.

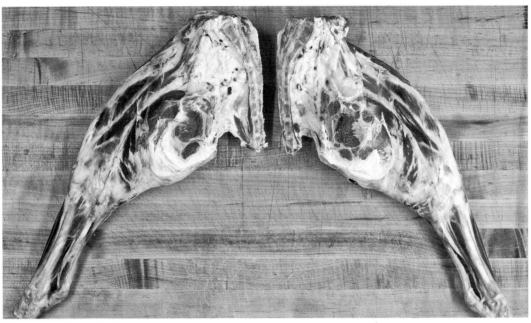

LEG SECTION, STEP 3: **The two leg sections.**

LEG SECTION, BONELESS LEG ROAST, STEP 1: **Cut behind the aitch bone on the pelvic bone, leaving as much meat as possible on the leg and keeping the bone nice and clean.**

LEG SECTION, BONELESS LEG ROAST, STEP 2: **Begin cutting the meat from the top of the aitch bone on the pelvic bone.**

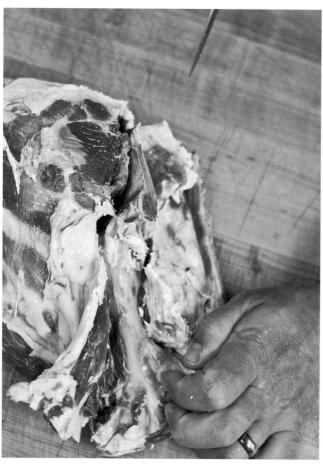

LEG SECTION, BONELESS LEG ROAST, STEP 3: Cut down toward the base of the pelvic bone, keeping the knife on the bone to leave behind as much meat as possible.

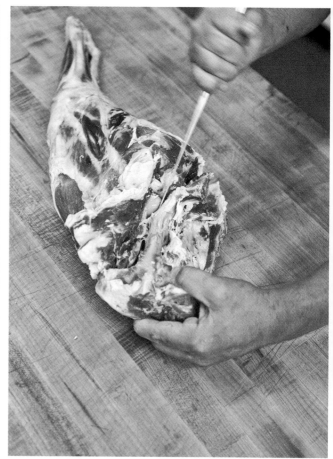

LEG SECTION, BONELESS LEG ROAST STEP 4: Free the joint from the leg bone by cutting through the ligament that connects the joint and pop it right out.

LEG SECTION, BONELESS LEG ROAST, STEP 5: **Work the bone back and forth and remove it as cleanly as possible.**

LEG SECTION, BONELESS LEG ROAST, STEP 6: **Fully removed pelvic bone—finish up the last cut to release it.**

LEG SECTION, BONELESS LEG ROAST, STEP 7: Following the muscle line, start cutting into the leg muscle from the top.

LEG SECTION, BONELESS LEG ROAST, STEP 8: Cut all the way down to one side of the bone, always keeping the knife on the bone.

LEG SECTION, BONELESS LEG ROAST, STEP 9: Cut cleanly and smoothly to remove the flesh from the bone.

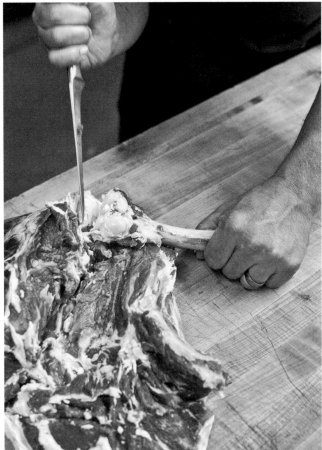

LEG SECTION, BONELESS LEG ROAST, STEP 10: Cut out the leg bone and the kneecap.

LEG SECTION, BONELESS LEG ROAST, STEP 11: Cut down at the base of the hind shank bone, into the joint.

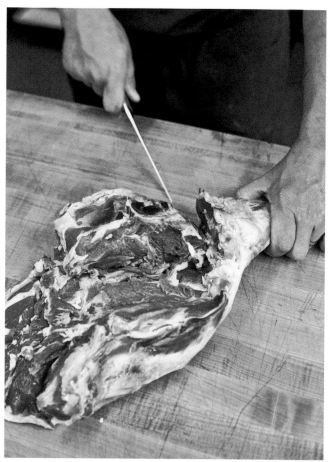

LEG SECTION, BONELESS LEG ROAST, STEP 12: Cut straight down below the base of the hind shank bone to release it.

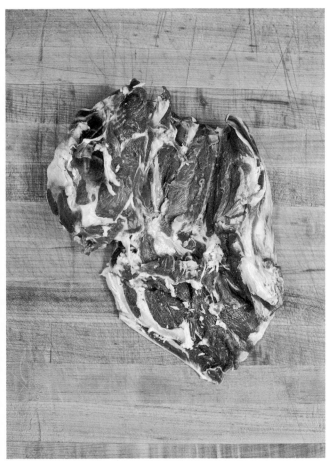

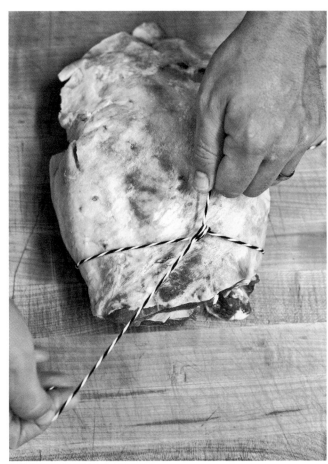

LEG SECTION, BONELESS LEG ROAST, STEP 13: **The boneless lamb leg (you could butterfly/flatten this for grilling). Just make sure to season it up nicely before you tie and roast.**

LEG SECTION, BONELESS LEG ROAST, STEP 14: **Begin to tie the roast—make the first tie at one end.**

LEG SECTION, BONELESS LEG ROAST, STEP 15: **Feed the butcher's twine underneath the first tie and make a tie every 2 inches/5 centimeters, feeding the string underneath and up for each tie.**

LEG SECTION, BONELESS LEG ROAST, STEP 16: **When you reach the other end of the roast, come back underneath and connect the butcher's twine to the first knot, then pull the string tight.**

LEG SECTION, BONELESS LEG ROAST, STEP 17: **The tied boneless lamb leg roast.**

LEG SECTION, LEG MUSCLES, STEP 1: Now, on the other leg: locate the line that indicates the seam between the muscles.

LEG SECTION, LEG MUSCLES, STEP 2: Make a series of small incisions down that line.

LEG SECTION, LEG MUSCLES, STEP 3: Pull back the first flap of muscle.

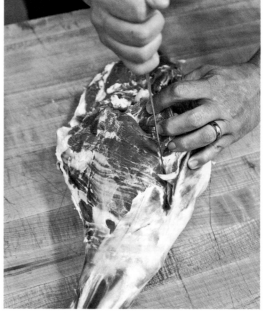

LEG SECTION, LEG MUSCLES, STEP 4: Continue cutting down along the line to release the first muscle. (Concentrate on only one muscle at a time.)

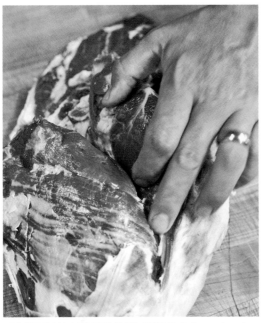

LEG SECTION, LEG MUSCLES, STEP 5: **Let your hands guide you as you release the muscle, making cuts into the connecting tissue as necessary.**

LEG SECTION, LEG MUSCLES, STEP 6: **Continue to pull apart the meat at the crease.**

LEG SECTION, LEG MUSCLES, STEP 7: **Pull the muscle up and cut the meat free all around and between the muscle's edges.**

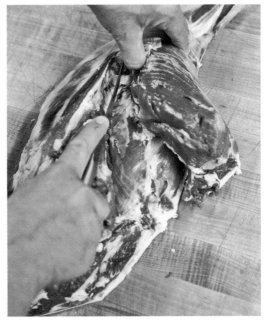

LEG SECTION, LEG MUSCLES, STEP 8: **Continue to cut until the muscle is free.**

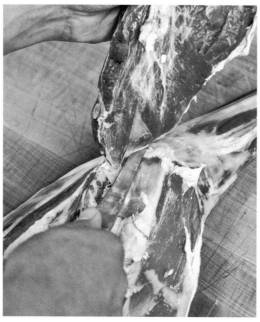

LEG SECTION, LEG MUSCLES, STEP 9: **Fully remove the first large muscle.**

LEG SECTION, LEG MUSCLES, STEP 10: **As soon as the first muscle is removed, you will see another line that defines the other muscles.**

LEG SECTION, LEG MUSCLES, STEP 11: **Now concentrate on removing the leg bone, keeping the knife as close to the bone as possible.**

LEG SECTION, LEG MUSCLES, STEP 12: **Fully free the leg bone at the base, leaving it connected to the shank. Now you can see the outlines of the other muscles.**

LEG SECTION, LEG MUSCLES, STEP 13: **Cut down, around, and between the muscles.**

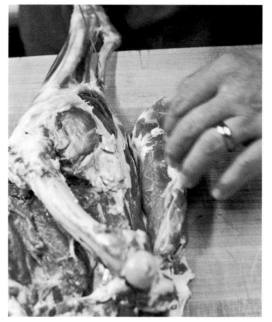

LEG SECTION, LEG MUSCLES, STEP 14: **Now pull off the next muscle.**

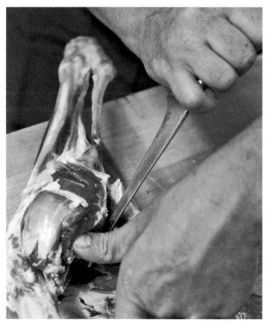

LEG SECTION, LEG MUSCLES, STEP 15: **When the base shank bone is fully exposed, cleanly remove it from the meat.**

LEG SECTION, LEG MUSCLES, STEP 16: **Peel off another one of the larger muscles.**

LEG SECTION, LEG MUSCLES, STEP 17: The five finished boneless lamb leg muscles.

Merguez Sausage

YIELD: 3.6 pounds/1.6 kilograms

These sausages are roller coasters of flavor. We have heat from the chile, great onion flavor, and plenty of fresh herbs. They are 100 percent lamb, even down to using sheep's casings, so they are a great alternative for someone who doesn't eat pork. Gently sear until medium-rare and serve on a bun with an herb salad and a light lemon vinaigrette. To make the sausages, you will need about 20 feet/6 meters of sheep's casings, which can be ordered in hanks (about 100 feet/30 meters) from a specialty butcher or on the Internet (see page 237).

Sheep's casings	About 20 feet/6 meters			
DRY INGREDIENTS				
Fine sea or kosher salt	2 tbsp	0.9 oz	26 g	1.6%
Fennel seeds, coarsely ground	½ tsp	0.1 oz	2 g	0.1%
Coriander seeds, coarsely ground	1½ tsp	0.1 oz	3 g	0.2%
Chipotle chile powder	¾ tsp	0.1 oz	2 g	0.1%
Ground cayenne pepper	1½ tsp	0.2 oz	5 g	0.3%
Ancho chile powder	1½ tsp	0.3 oz	7 g	0.4%
Cumin seeds, coarsely ground	1½ tsp	0.1 oz	3 g	0.2%
Caraway seeds, coarsely ground	¾ tsp	0.1 oz	2 g	0.1%
Dried red chile flakes	¼ tsp	0.1 oz	2 g	0.1%
Sweet paprika	1½ tsp	0.2 oz	5 g	0.3%
Granulated sugar	1 tsp	0.2 oz	5 g	0.3%
Freshly ground black pepper	1 tsp	0.1 oz	3 g	0.2%
Curing salt (pink salt or sodium nitrate)	½ tsp	0.1 oz	2 g	0.1%

cont'd

MEAT				
Boneless lamb neck (see page 94), very cold	1 whole	3.1 lb	1,409 g	85.6%
WET MIX				
Jalapeño, finely chopped	¾ small	0.4 oz	12 g	0.8%
Finely chopped red onion	2 tbsp	0.9 oz	25 g	1.5%
Minced garlic	½ tsp	0.2 oz	4 g	0.3%
Water	⅓ cup	3 oz	86 g	5.2%
Fresh oregano, finely chopped	2 sprigs	0.1 oz	3 g	0.2%
Finely chopped scallion	2 tbsp	0.7 oz	21 g	1.2%
Fresh cilantro, finely chopped	11 sprigs	0.7 oz	21 g	1.2%

1. Read the General Sausage-Making Tips (see page 17).

2. The night before: Soak the sheep's casings in a bowl of cold water; refrigerate overnight.

3. Assemble all of the dry ingredients in one container and all of the wet mix (except the fresh herbs) in another. (This step does not need to be done the night before, but it's crucial that it be completed before you start grinding the meat.)

4. The next day: Untangle the casings, and begin to open them to make the stuffing process easier. Hold one end of each piece of casing up to the nozzle of the faucet and support it with your other hand. Gently turn on the water and let it run through the casings to check for holes. If there are any holes in the casings, cut out the pieces with holes. Hold the casings in a bowl of ice water or refrigerate until stuffing time.

5. With a sharp boning knife or your knife of choice, remove the meat and fat from the lamb neck. Open-freeze the meat, uncovered, for 30 to 60 minutes, until the surface of the meat is crunchy to the touch and the interior is very cold, but not frozen.

6. Cut the meat into 1-inch-/2.5-centimeter-square cubes or a size slightly smaller than the opening of the meat grinder. Open-freeze the meat again, uncovered, for 30 to 60 minutes, until the surface of the meat is crunchy to the touch and the interior is very cold, but not frozen.

7. When you are ready to grind, prepare a perfectly clean and chilled meat grinder for grinding, and fit it with the medium plate. Start the auger, and, without using the supplied pusher, let the auger gently grab each cube of meat and bring it forward toward the blade and through the grinding plate. Continue grinding until all of the meat has been processed. Place it in a clean, cold non-reactive bowl or tub and again open-freeze, uncovered, for 30 to 60 minutes, until the surface of the meat is crunchy to the touch and the interior is very cold, but not frozen.

8. In a medium nonreactive bowl, combine the wet mix and dry ingredients, stir in the herbs, and whisk together until completely blended and the dry ingredients have dissolved. I call this the "slurry."

9. In a large, wide basin or bowl that will give you plenty of room to mix the meat and seasonings, combine the cold meat with the slurry. Roll up your sleeves and, with perfectly clean hands, begin kneading and turning the mixture as you would a large quantity of bread dough. Eventually, you will begin to notice that the mixture has acquired a somewhat creamy texture. This is caused by the warmth of your hands and is a sign that you have finished mixing. Spoon out a few tablespoons of the mixture, and return the remainder to the refrigerator.

10. In a nonstick skillet over medium heat, lightly fry a test portion of sausage mixture until cooked through but not caramelized (which would change the flavor profile). Taste for seasoning. Based on this taste test, you can adjust the salt in the main portion of sausage, if desired.

11. Prepare a perfectly clean and chilled sausage stuffer and place the water-filled bowl of casings next to it. You will also need a landing surface of clean trays or parchment paper–lined baking sheets for your finished sausages.

12. Load the sausage mixture into the canister of the sausage stuffer, compacting it very lightly with a spatula to be sure there are no air pockets. Replace the lid.

13. Thread a length of casing all the way onto the stuffing horn and start cranking just enough to move a little of the ground meat mixture into the casing. As soon as you can see the meat poking through the nose of the stuffer, stop and crank backward slightly to halt the forward movement. Pinch the casing where the meat starts (to extrude all the air), and tie into a knot. Now start cranking again with one hand while you support the emerging sausage with the other. Move the casing out slowly to allow it to fill fully but not too tightly, so that there will be some give in the sausage when it comes time to tie the links. When you get close to the end, leave 6 inches/15 centimeters of unstuffed casing and stop cranking.

14. Go back to the original knot and measure 6 inches/15 centimeters of sausage. Pinch the sausage gently to form your first link, and twist forward for about seven rotations. Move another 6 inches/15 centimeters down the sausage, and this time pinch firmly and twist backward. Repeat this process every 6 inches/15 centimeters, alternating forward and backward, until you reach the open end of the casing. Twist the open end right at the last bit of sausage to seal off the whole coil, and then tie a knot.

15. Ideally, hang the sausage overnight in a refrigerator, or refrigerate on parchment paper–lined baking sheets, covered with plastic wrap, to allow the casing to form fully to the meat, and the sausage to settle. (Or, if desired, you can cook the sausages right away.) The next day, cut in between each link and cook as desired.

Roasted Leg of Lamb with Rosemary Pesto

SERVES 10

I can see the point of roasting meat on a rack, but for this recipe I prefer
to roast the meat on a bed of vegetables, which turns into overcooked
goodness, like baba ghanoush—a salty-herbal-sweet pillowy layer
of flavor. The initial high-heat searing creates a great, crispy crust
on the outside, while the meat inside remains meltingly tender.
If you like, you can grill the roast over hardwood, which produces
a tasty and perfect special-occasion roast.

ROSEMARY PESTO				
Small, tender fresh rosemary sprigs, loosely packed	1 cup	1.1 oz	30 g	0.08%
Garlic cloves, peeled	6 large	1.6 oz	46 g	1.35%
Red onion	¼ large	3.4 oz	95 g	2.6%
Extra-virgin olive oil	⅔ cup	5.4 oz	152 g	4.1%
Fine sea salt	2 tsp	0.43 oz	12 g	0.3%
Freshly ground black pepper	1 tsp	0.1 oz	2 g	0.07%
MEAT				
Boneless leg of lamb (see page 127), at room temperature	1 whole	5.7 lb	2,600 g	70.8%
VEGETABLES				
Japanese eggplant, cut into large dice	1	10.6 oz	227 g	8.5%
Sweet potato, peeled, quartered lengthwise and cut into large dice	1	10.6 oz	227 g	8.5%
Fine sea salt	1 tsp	0.22 oz	6 g	1.6%

Extra-virgin olive oil	¼ cup	2.1 oz	60 g	0.2%
Rosemary sprigs, halved crosswise	3 large	0.5 oz	15 g	0.4%
Garlic, loose papery skin removed and root trimmed off	1 whole head	1.9 oz	55 g	1.5%
Grapeseed oil, for searing	as needed			

1. In a Vita-Mix or bar blender, make the Rosemary Pesto: Puree the rosemary, garlic, onion, olive oil, salt, and pepper until smooth (it's okay if it's a little chunky). You can also do this in a food processor, as long as you scrape down the sides and process until smooth and blended.

2. Lay out the lamb leg on a work surface and butterfly/cut into the large muscles to create a larger area for the pesto to penetrate (see page 125). Massage the pesto into both sides, working it into all the nooks and crannies with your fingers. Cover with plastic wrap and refrigerate overnight. Remove from the refrigerator 1½ hours before you plan to begin roasting.

3. Toss together the eggplant, sweet potato, salt, olive oil, and rosemary. Trim off the top of the head of garlic to expose the cloves and halve it through the top. Toss the garlic halves with the vegetables. Spread the mixture in a small roasting pan just slightly larger than the lamb itself so that all the vegetables will benefit from the drippings.

4. Preheat the oven to 275°F/135°C/gas 1. Tie the room temperature lamb back together into the shape of the leg, tying once lengthwise and two or three times crosswise (see pages 125–127). In a pan large enough to accommodate the lamb, heat a little grapeseed oil until very hot but not smoking and sear the roast on all sides until the surfaces are brown and crusty. Lift out the seared lamb and place it on top of the vegetables in the roasting pan. Insert a probe thermometer in the center of the meat and place the controls outside the oven. If your thermometer has an alarm, set it for 127°F/52°C.

5. Slow-roast the lamb until the internal temperature reaches 127°F/52°C, which will be 2 hours and 15 minutes to 2½ hours. Loosely, but completely, cover the lamb with two sheets of aluminum foil and allow it to rest for about 20 minutes. Slice and enjoy with the softened vegetables.

Braised Lamb Shanks with Curry

························

SERVES 4

Lamb shanks are super-tasty when cooked slow-and-low, as in this braise, and they definitely make a hearty meal that will warm your taste buds. This dish has a full range of flavors going on: The curry and chile bring in some nice heat, the carrot helps balance the heat with its own sweetness, and the crème fraîche and final squeeze of lime juice help cut through and brighten up a typically hearty wintertime dish, making it great during any season, not just fall or winter.
I like to serve it family-style, over basmati rice. If you use Vadovan curry, use double the amount called for here.

MEAT				
Lamb shanks (see page 98), about 1 lb/455 g each	4 large	4 lb	1,816 g	35.9%
Fine sea salt	2 tsp	0.05 oz	13 g	0.3%
Coarsely ground black pepper	as needed			
Grapeseed oil	as needed			
VEGETABLES AND BROTH				
Red onion, cut into large chunks	1	13.4 oz	380 g	8%
Carrot, peeled and cut into large chunks	1 medium	3.5 oz	98 g	1.9%
Fingerling potatoes, washed but unpeeled	7 small	17.6 oz	500 g	10%
Ginger, peeled and cut into large chunks	1-inch/2.5-cm slice	1.4 oz	40 g	0.8%
Dried Thai or bird's-eye chiles	2 whole			
Madras curry powder	1½ tsp	0.4 oz	10 g	0.2%
Garlic, loose papery skin removed and top one-third trimmed off	1 whole head	1.9 oz	55 g	1.1%
Carrot juice	2 cups	16.9 oz	480 g	9.5%
Freshly squeezed orange juice	¼ cup	2.1 oz	60 g	0.1%

Water or light lamb or vegetable stock	6¼ cups	49.7 oz	1,410 g	29%
Fine sea salt	1 tsp	0.1 oz	5 g	0.12%
Orange zest (about one-third of an orange)	3 strips	0.1 oz	3 g	0.1%
Slivered fresh mint leaves	1¼ cups	1 oz	25 g	0.48%
Slivered fresh cilantro leaves	2 cups	1.1 oz	30 g	0.6%
Slivered scallion greens	1 bunch	1.4 oz	40 g	0.1%
Freshly grated ginger (on a microplane)	½ tsp	0.1 oz	4 g	0.1%
Fresh lime	1 whole			0.5%
Crème fraîche or Greek yogurt	¼ cup	2.1 oz	60 g	1.2%

1. Season the lamb shanks generously with the salt and lightly with pepper. Warm a generous amount of grapeseed oil (1 to 2 tablespoons, depending on the size of the pan) in a large ovenproof braising pan or Dutch oven over medium-high heat. Sear the shanks until nicely colored on all sides.

2. Add the onion, carrot, potatoes, ginger, chiles, and curry to the pan over the shanks. Halve the head of garlic lengthwise and place the halves, cut-sides down, directly on the hot surface of the pan. Turn the lamb shanks on their narrow sides and, using the vegetables as support, sear the thin edges.

3. Preheat the oven to 325°F/165°C/gas 3. Add the carrot juice, orange juice, water, salt, and orange zest to the shanks and bring to a boil. Cover the pan with a large sheet of parchment paper, pressing it down so that it touches the surface of the liquid and extends up the sides and over the rim of the pan. Place the lid over the paper and transfer the pan to the oven. Braise until the lamb is falling off the bones, about 2 hours. (Check every 30 to 45 minutes and lower the oven heat

by a few degrees if the liquid is simmering briskly—it should barely quiver.)

4. Remove and discard the parchment. Transfer the shanks, still on the bone, to a platter and strain the braising liquid. From the strainer, remove the potatoes and garlic pieces and reserve; discard the remaining solids. Squeeze the softened garlic into the braising liquid and reduce the juices by about half over medium-high heat.

5. Cut the potatoes into 1-inch/2.5-centimeter rounds and return them, with the lamb shanks, to the braising liquid. Warm through for 5 minutes, then stir in the mint, cilantro, scallion greens, and ginger. Squeeze in all the juice from the lime. Remove the pan from the heat. Taste for seasoning.

6. Serve each shank in a wide bowl with plenty of potatoes and the garlicky braising liquid. Top each serving with a dollop of crème fraîche and enjoy.

Lamb Loin Wrapped in Its Belly

SERVES 6

This is just like a porchetta, only smaller (and, of course, it's lamb not pork). The cut I use here is the completely boned-out rack of lamb but with a much longer piece of the belly attached than usual (see page 107). You could make this with a rack from your local butcher (as long as the bones have not been frenched, as roasts sometimes are), and it will work quite well. Just ask your butcher to bone it out and keep the belly on.

One of my favorite ways to serve this—I call it "lambchetta"—is on a fresh tortilla with roasted eggplant and "marrownaise"—my version of homemade mayonnaise, in which I use rendered marrow fat in place of some of the oil.

Lamb loin (see page 107), connected to the belly and completely boned	1 whole	31.3 oz	888 g	92.3%
Fine sea salt	2¼ tsp	0.5 oz	14 g	1.5%
Freshly ground black pepper	as needed			
Extra-virgin olive oil	2½ tbsp	0.4 oz	12 g	1.2%
Slivered flat-leaf parsley leaves	3 tbsp	0.1 oz	3 g	0.3%
Slivered fresh cilantro leaves	3 tbsp	0.1 oz	3 g	0.3%
Slivered fresh mint leaves	3 tbsp	0.1 oz	3 g	0.3%
Slivered fresh basil leaves	3 tbsp	0.1 oz	3 g	0.3%
Dried red chile flakes	1½ tsp	0.1 oz	2 g	0.2%
Coarsely ground black pepper	¾ tsp	0.1 oz	2 g	0.2%
Ground coriander seeds	¾ tsp	0.1 oz	2 g	0.2%
Paprika	1½ tsp	0.1 oz	3 g	0.3%
Freshly grated lemon zest (on a microplane)	¼ tsp	0.04 oz	1 g	0.1%
Finely chopped shallot	3 tsp	0.8 oz	24 g	2.5%
Garlic cloves, thinly sliced	2 large	0.1 oz	3 g	0.3%

1. Season all surfaces of the lamb generously with the salt and lightly with pepper. In a medium bowl, combine 1 tablespoon of the olive oil with the parsley, cilantro, mint, basil, chile flakes, pepper, coriander seeds, paprika, lemon zest, shallot, and garlic and stir until well blended. Place the loin and belly, fat-side down, on a work surface. Spread the stuffing evenly over the meat and belly, leaving a ¾-inch/2-centimeter border around the edge.

2. Begin rolling up the cylinder firmly, beginning with the eye and finishing with the end of the belly meat. Tie the cylinder once in the center, and then make the second tie 1½ inches/ 4 centimeters below the center tie, the third tie 1½ inches/4 centimeters above the center tie, alternating back and forth to maintain an even thickness until you reach each end.

3. Preheat the oven to 275°F/135°C/gas 1. Brush the outside of the roast with 1½ teaspoons of the olive oil and season again with salt and pepper. In a roasting pan large enough to accommodate the lamb, heat the remaining 1 tablespoon olive oil until very hot but not smoking. Sear the lamb until the outside is caramelized and the fat has been rendered. Remove the lamb and place a roasting rack in the pan. Place the lamb on the rack and roast until the internal temperature at the center reaches 135°F/57°C, approximately 1 hour.

4. Tent the roast loosely with aluminum foil and allow it to rest for 5 minutes, then snip and remove the strings. Slice into 1-inch/2.5-centimeter-thick medallions and enjoy.

Lamb Leg Muscles Wrapped in Bacon

SERVES 4

Depending on which leg muscles you choose, the muscles may be small or large, thick or thin. By butterflying the muscles individually, you turn them into rolls that can easily be cooked to an even temperature. The bacon protects the lean meat from drying out and keeps it nice and juicy. Rolls with a wider diameter will take longer, of course, but you should always let the internal temperature—not the time—be your guide to doneness. You can substitute six slices of uncured belly meat for the bacon; it should be thinly sliced but not paper thin, more like business-card thin. Note: Even though string is pretty tasty, make sure that you count the number of strings when you tie, so that you can count them again when you remove them.

Lamb leg muscles (see page 132), separated with the natural seams (about 2)	1½ lb	24.7 oz	700 g	62.7%
Fine sea salt	1 tsp	0.2 oz	6 g	0.5%
Coarsely ground black pepper	as needed			
Grape leaves in brine, drained and minced	2 tbsp	1.1 oz	30 g	2.7%
Fire-roasted jarred red peppers, drained and minced	1½ tbsp	2.1 oz	60 g	5.4%
Minced flat-leaf parsley	2 tsp	0.1 oz	4 g	0.4%
Coarsely chopped toasted almonds	2 tsp	0.1 oz	2 g	0.2%
Pitted black olives, chopped	6 small	0.5 oz	14 g	1.3%
Freshly grated lemon zest (on a microplane)	¼ tsp	0.04 oz	1 g	0.04%
Cured bacon, about 1¾ in/4.5 cm wide and as thick as a thin coin	6 slices per piece of lamb	10.6 oz	300 g	26.8%
Grapeseed oil or clarified butter	as needed			

1. Trim off any sinew or fat on the lamb leg pieces that looks inedible, but don't overdo the trimming; leave a little bit of fat.

2. Now butterfly the meat. Holding a muscle with the short end closest to you and steadying it with the palm of your left hand (or right hand, if you're left-handed), slice into the long, right side halfway between the top and bottom, slicing with the grain almost but not all the way through toward the left, longer end. Open out the meat and press gently to flatten it. You should have a relatively even and flat piece of meat, double the width of the piece you started with.

3. Mix together the grape leaves, roasted red peppers, parsley, almonds, olives, and lemon zest. Season the meat inside and out with the salt and pepper. Make a 1-inch/2.5-centimeter-wide line of stuffing along the seam of each butterflied muscle, and fold one side over the top as if you were closing a book.

4. Lay out a large sheet of plastic wrap on the work surface for each piece of meat you will be rolling. Arrange six bacon strips parallel to each other, and very slightly overlapping, in a large square in the center of each piece of plastic, placing them right next to each other and alternating their direction to compensate for the fact that one end is usually thinner than the other.

5. Place one of the stuffed muscles on the lower third of the bacon sheet, perpendicular to the bacon's direction. Using the long end of plastic closest to you as a helper, begin rolling up the muscle inside the bacon, pushing and compacting to make a firm, round roll but making sure to keep the plastic on the outside only. Roll up tightly, as if you were making a roll of compound butter, making it as firm as possible and about 2½ inches/ 6 centimeters in diameter. Twist the ends tightly and refrigerate for 5 hours, to help the bacon stick to the meat.

6. Carefully remove the muscles from the plastic and tie every 1 inch/2.5 centimeters, keeping the knots on the side of the roll exactly *opposite* from where the short ends of the bacon meet or overlap. Season the outside lightly with salt and pepper.

7. Preheat the oven to 275°F/135°C/gas 1. In a large, ovenproof skillet, warm a generous amount of grapeseed oil over medium-high heat until very hot but not smoking. Place the roll seam-side down first (knot-side up) so that it does not unravel. Lightly sear each roll, turning it occasionally for even cooking, until light golden brown on all sides. (If you cook the rolls too much at this point, the bacon will harden, crack, and pull away from the meat.)

8. Insert a probe into the center of the smaller piece of meat and slow-roast it until the internal temperature at the center reaches 135°F°/57°C, about 30 minutes. Remove the smaller piece from the oven, insert the probe into the larger piece, and continue cooking to 135°F/57°C. Cover both rolls loosely with aluminum foil and allow them to rest for 8 to 10 minutes. Snip and remove the pieces of string, then cut the rolls crosswise into ½-inch-/12-millimeter-thick slices with a sharp knife, and enjoy.

CHAPTER THREE

PORK

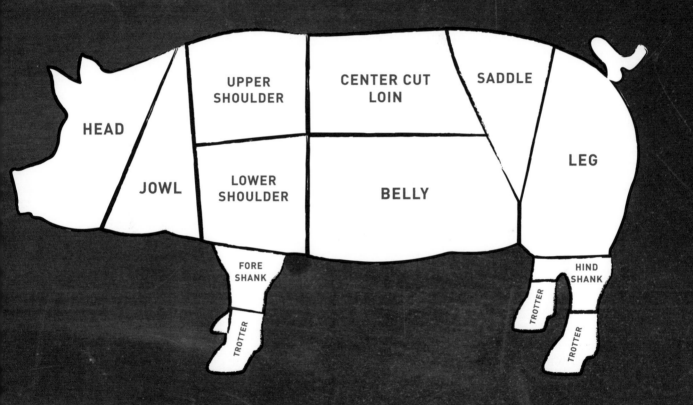

Pigs can range in size from 20 pounds/9 kilos, for a suckling, to 250-plus pounds/113-plus kilos, which is considered market size. But at home it's *much* easier to work with an 80-pound/36-kilo animal, as we do here. The pig is smaller than the steer, so we begin by separating it into manageable sections, then breaking those sections into two halves. Let your fingers be your cutting guide: feel the natural seams in the muscles before making your marks with a knife. Then, the road map is ready for you.

I encourage you to make a **head roast**. It's a fun, ambitious, and challenging project, because you have to take the whole face off the skull. There are a lot of detailed cuts if you want to get all of the meat out of the dips and crevices on the skull. Keeping the jowl attached, start from the bottom and work your way up. Using the skin from the shoulder, place pieces of skin over the empty eye sockets to seal off the openings. Stuff it with some seasoned, diced meat from the upper shoulder. Then suture it all up, using a big piece of skin from the shoulder to hold the meat inside where the neck opening was.

After the **neck**, the **tongue** is probably my favorite part of any animal; it's so tasty and full of goodness. Salt for two days, and braise for a long time—no need to peel off the skin, as you must with beef—then let it cool down. Slice thinly for tongue salad or sandwiches. The **ears** have great texture. The cartilage in the ear is not really edible, but when cooked for a long time, cooled, and thinly sliced, you can deep-fry the ears for a salad. They make a great crunchy topping— better than croutons!

I like to salt the **trotters** for a couple days and then braise them. Once cooled to room temperature, the cooked skin and meat can be pulled off and put into a terrine or used for sausage.

There is a lot of great marbling in the **shoulder**—this is a fantastic section, with a good mix of meat to fat (about 80 percent meat and 20 percent fat) and not a lot of bone. Boneless shoulder meat is great for confit or sausage. And it's accessible: You don't have to get a whole shoulder section. This is a great place for the beginner to start, because there are fewer bones to deal with, and it's a forgiving cut to cook; no matter what you do, you're going to end up with delicious meat.

A **picnic ham** is a squared-off cut from the lower shoulder that has nothing to do with the leg (also called a ham). There is far more fat in this section of the animal than in the leg. You can roast it with the skin on, or brine and smoke it.

The **bone-in chop** is a great cut for grilling, I always baste it to keep it moist. This part of the animal has a little less fat, so use a marinade (water, salt, mustard, vinegar, brown sugar) and dip the chops in it every time you flip them. This slows the cooking process a little but gives you more flavor, plus great smoke when the drippings hit the coals.

Everybody loves **pork belly**. Here, we prepare the belly in two ways, one with the rib meat on and one with the rib meat off. (The rib meat is usually included on the St. Louis ribs.) Regardless, rib meat on or off, this whole section is great roasted till the skin is crispy, and served with greens. Standing room only.

I love the **saddle**. It's the gem of the whole animal, especially if you're cooking for a dinner party. The **tenderloin** is awkwardly placed between two sections and must be pulled out of position in order to avoid cutting into it. The belly here is a little fattier than it is in the center of the loin and can be used wherever you need extra flavor and fat, such as in sausage.

The tenderloin is not one of my favorite cuts, because it is so lean, but if you leave the fat on—and don't overcook it—you'll have more flavor, especially with a young animal. You have to add fat aggressively to keep it from drying out. I like to grill it quickly just to add grill marks, then finish cooking slow-and-low by poaching it in butter with a bunch of herbs. (Yes, actually immersing it in herb butter.)

The **porterhouse** section is especially accessible and is easy to order from your butcher. It makes an impressive huge roast (or two- or three-chop section). Cook fat-side down, first over high, direct heat and then finish slow-and-low.

The **leg** is a challenging part of the animal because it has a lot of lean meat; the animal walked around using these muscles all the time. When I prepare a **whole leg**, I brine it first for at least four days, and then I inject more brine into the bone area. The **cowboy steaks** are great by themselves but also perfect for making your own tasso.

When removing the **skin**, make sure that you leave as much fat as possible on the meat and as little as possible on the skin. This makes it possible to make the skin into cracklings and ensures the meat will stay moist and juicy when cooked. If you want to, you can always remove the fat after cooking.

Pork is very rich meat and is so tasty. When preparing it, I cut that richness with some brightness, in the form of vinegar and fresh greens; balance on the palate makes happy bellies.

Always follow your stomach when making major decisions in the kitchen!

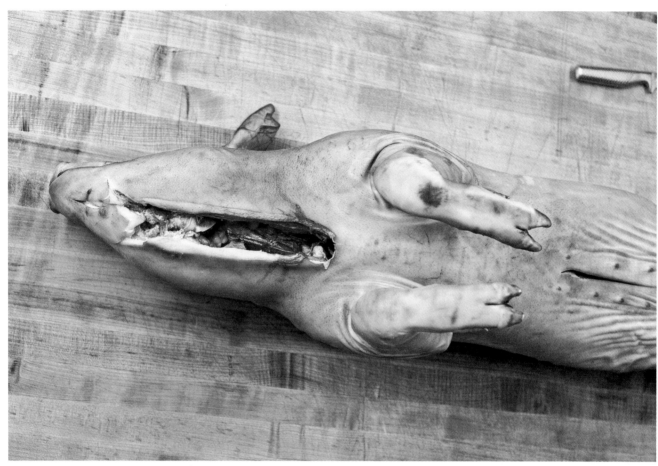

WHOLE PIG, STEP 1: A roaster-size pig (about 80 pounds/36 kilos) is a nice manageable size—easier to work with than a market-size pig of 250 pounds/113 kilos, which can be difficult to maneuver when whole.

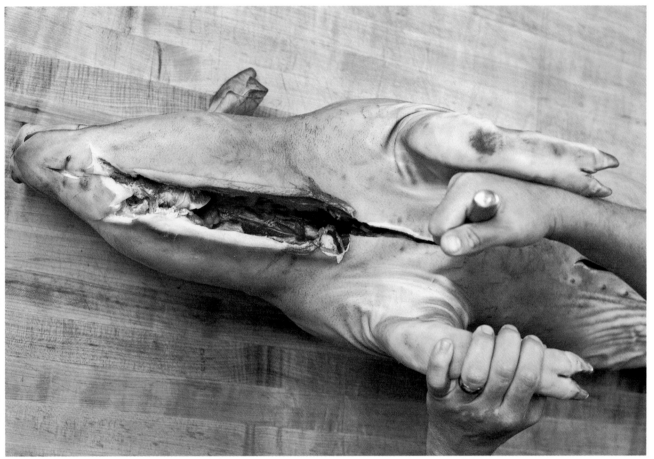

WHOLE PIG, STEP 2: Make the first cut through the center of the breastbone. Saw if you need to, but with a smaller pig like this, a sharp knife applied with firm pressure works just as well.

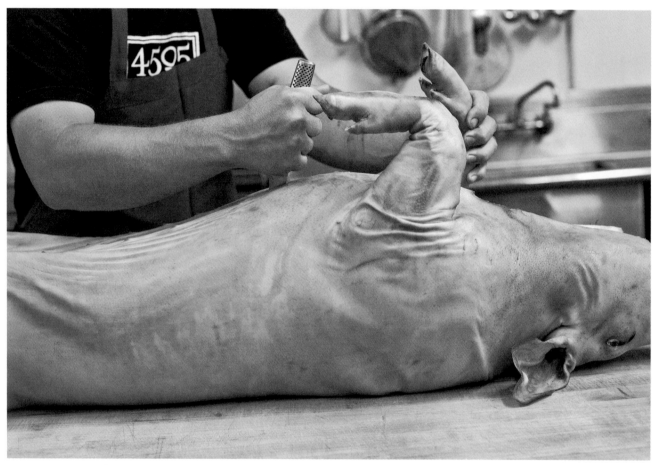

WHOLE PIG, STEP 3: **Hold a leg for leverage to keep the animal steady while cutting through the breastbone.**

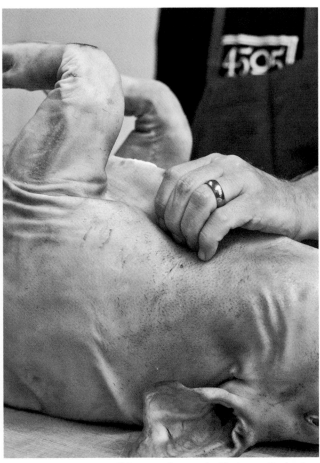 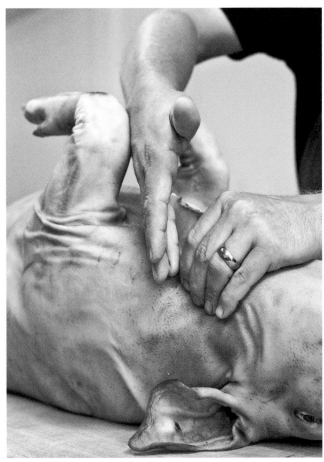

HEAD SECTION, STEP 1: Using your fingers, feel between the jowl and shoulder to find where the muscles begin and end. Pinching the flesh will help you locate the thinnest part.

HEAD SECTION, STEP 2: Here, between the shoulder and the jowl, is the thinnest part of the neck; it's just skin. This is the natural crease, where you will make your first cut.

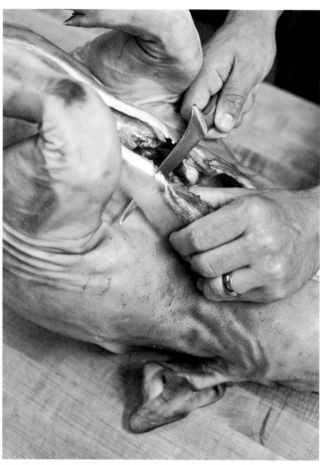

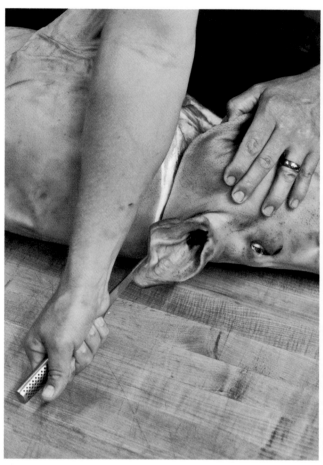

HEAD SECTION, STEP 3: Poke the knife through the thinnest part, where there is no muscle, to make a starter cut. Cut carefully toward yourself.

HEAD SECTION, STEP 4: Finish the incision with the knife, cutting down all the way through the meat to the neck bone.

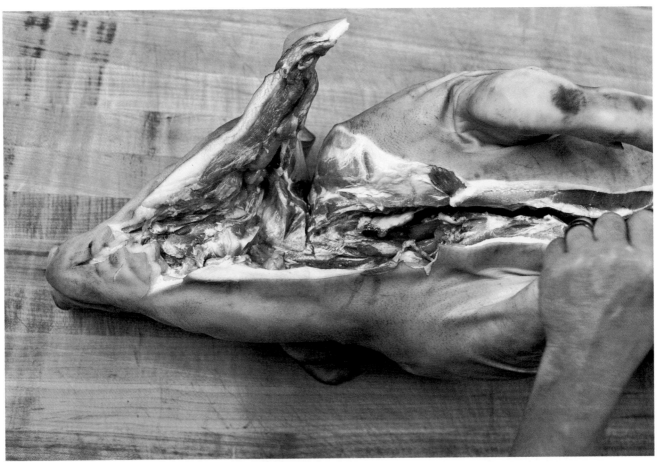

HEAD SECTION, STEP 5: See how the jowl has pulled away from the shoulder without your having cut through either of the muscles.

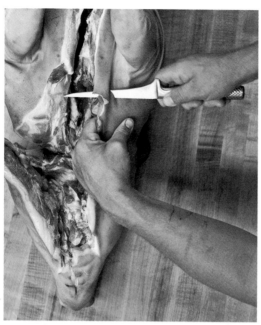

HEAD SECTION, STEP 6: Repeat the process on the other side.

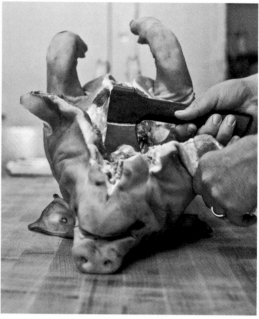

HEAD SECTION, STEP 7: Chop through the neck bone with a hatchet, cleaver, or bone saw—whichever you are most comfortable using.

HEAD SECTION, STEP 8: Using the knife again, cut through any meat that's still attached. General rule: hatchet, cleaver, and saw for bone; knife for flesh.

HEAD SECTION, STEP 9: Work nice and clean; make sure you always clean up any bone shards or bone dust before moving on to the next step.

HEAD SECTION, STEP 10: Hey, buddy!

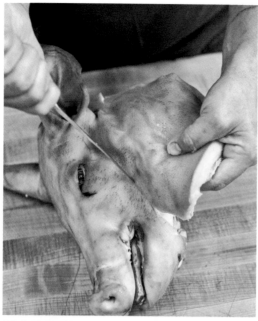

HEAD SECTION, JOWL, STEP 1: **Make an incision right at the cheekbone, below the eye socket; begin peeling off the jowl.**

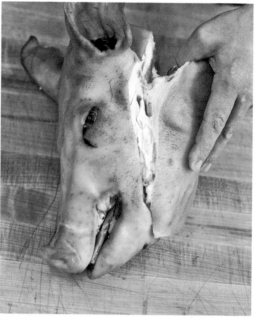

HEAD SECTION, JOWL, STEP 2: **Now, exposing the jawbone, plan your next cut.**

HEAD SECTION, JOWL, STEP 3: **Continue working close to the bone to finish the cut, leaving as much meat on the jowl as you can.**

HEAD SECTION, JOWL, STEP 4: **The two finished jowls.**

HEAD SECTION, TONGUE, STEP 1: **Pull up the base of the tongue and cut the bottom of the tongue free with your knife.**

HEAD SECTION, TONGUE, STEP 2: **Pull out the tongue.**

HEAD SECTION, TONGUE, STEP 3: **Fully free the tongue from the head.**

HEAD SECTION, TONGUE, STEP 4: **The finished tongue.**

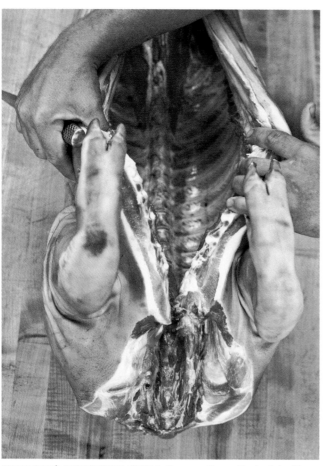

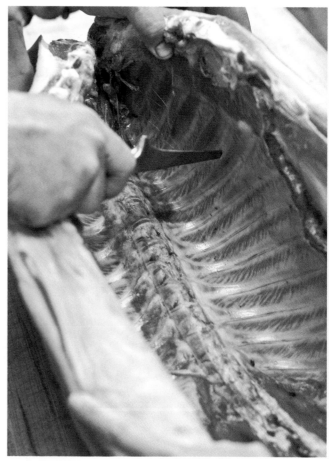

SHOULDER/LOIN SECTION, STEP 1: Open up the cavity, look inside, and make a game plan; identify your next incision.

SHOULDER/LOIN SECTION, STEP 2: Make the first cut between the 4th and 5th ribs, to separate the belly section from the shoulder section. (If you cut between the 3rd and 4th ribs, you'll run into the thick part of the shoulder blade.)

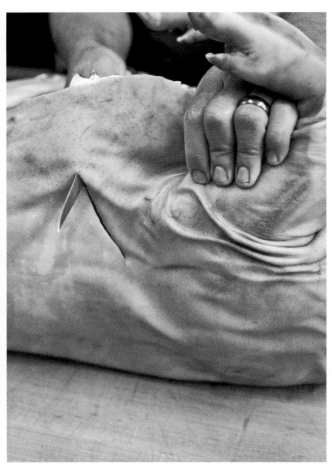

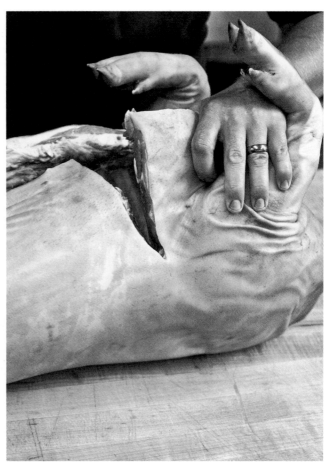

SHOULDER/LOIN SECTION, STEP 3: Push the knife through the flesh, and begin cutting up at an angle between the rib bones, following the angle of the bones.

SHOULDER/LOIN SECTION, STEP 4: After making the first incision, gently pull apart the ribs to see and plan your next cut.

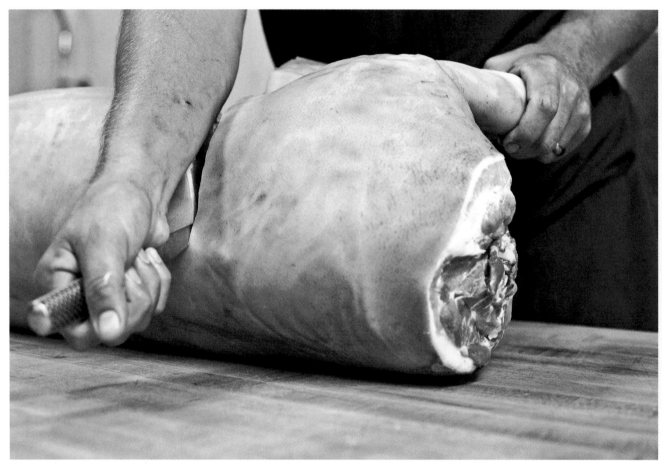

SHOULDER/LOIN SECTION, STEP 5: Continue all the way around the spine to free the meat.

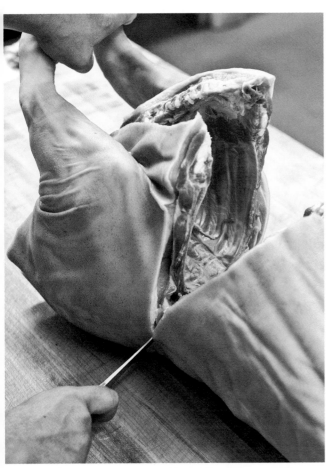

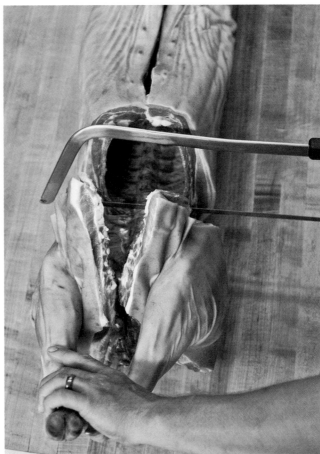

SHOULDER/LOIN SECTION, STEP 6: **Repeat the process—cutting through flesh, not the bone—on the other side.**

SHOULDER/LOIN SECTION, STEP 7: **Come back with the saw to cut through the spine.**

SHOULDER/LOIN SECTION, STEP 8: Come back with the knife and a wet towel to clean up the bone dust after sawing.

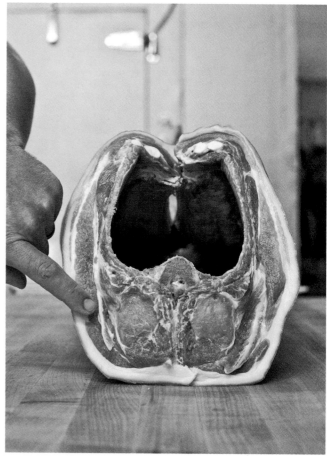

SHOULDER SECTION, STEP 1: Notice the location of the tip of the shoulder blade; this is what you want to see. This allows us to maximize the amount of bone-in pork chops, and still have plenty of tasty meat on the shoulder.

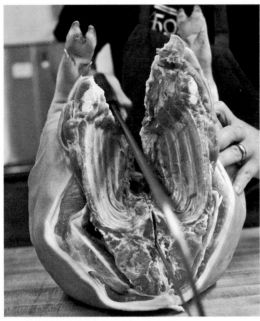

SHOULDER SECTION, STEP 2: **Saw directly down the spine.**

SHOULDER SECTION, STEP 3: **The vertebrae curves down at the neck; therefore, saw at an angle to finish splitting the spine.**

SHOULDER SECTION, STEP 4: **Come back with the knife to finish cutting through the meat.**

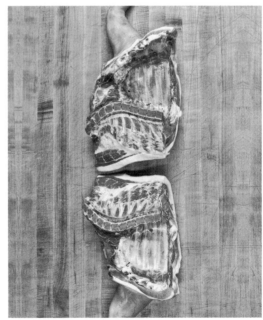

SHOULDER SECTION, STEP 5: **The full shoulder section, bone and trotter intact.**

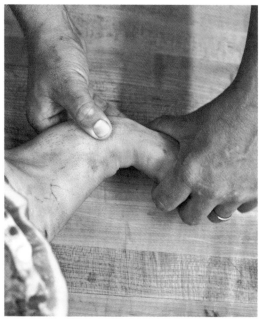

SHOULDER SECTION, TROTTER, STEP 1: "Hello, nice to meat you." Use your thumb to feel where the joint flexes, or, in other words, find the soft spot.

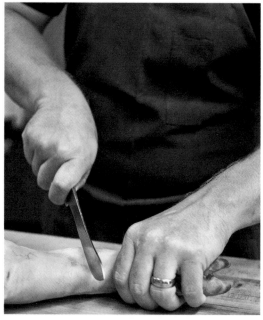

SHOULDER SECTION, TROTTER, STEP 2: Mark the soft spot.

SHOULDER SECTION, TROTTER, STEP 3: Make the first incision. Moving the joint back and forth, work the knife downward, following the path of least resistance down through the joint.

SHOULDER SECTION, TROTTER, STEP 4: Pop up the joint to help release the leg and finish cutting through the rest of the skin.

SHOULDER SECTION, SHOULDER ROAST, STEP 1:
Cut through the soft cartilage between the ribs and the breastplate.

SHOULDER SECTION, SHOULDER ROAST, STEP 2:
Continue to cut all the way through the cartilage.

SHOULDER SECTION, SHOULDER ROAST, STEP 3:
Come back and fully remove the breastplate.

SHOULDER SECTION, SHOULDER ROAST, STEP 4:
Saw through the vertebrae before the first rib.

SHOULDER SECTION, SHOULDER ROAST, STEP 5:
When removing any bone from the meat, always keep one side of the knife on the bone. This will help you maximize your yield of meat.

SHOULDER SECTION, SHOULDER ROAST, STEP 6:
Come back and remove the neck bone with the knife.

SHOULDER SECTION, SHOULDER ROAST, STEP 7:
Saw through the spine to free up the feather bone, then remove it with the knife.

SHOULDER SECTION, SHOULDER ROAST, STEP 8:
Semi-boneless shoulder section.

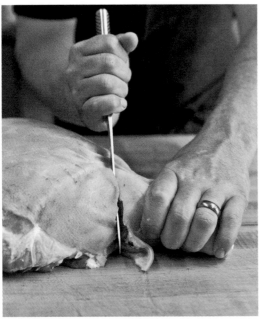

SHOULDER SECTION, SHOULDER ROAST, STEP 9:
Remove the foreshank by cutting toward the arm joint at the base of the arm bone. Moving the foreshank back and forth, cut it free from the arm bone.

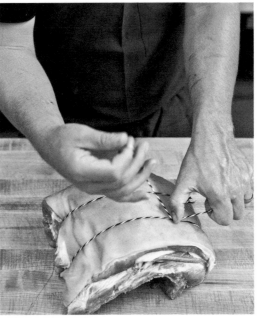

SHOULDER SECTION, SHOULDER ROAST, STEP 10:
Tie up the shoulder roast.

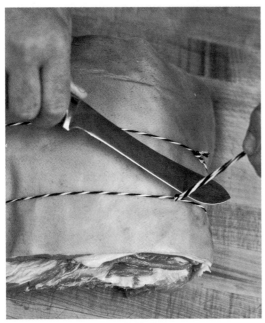

SHOULDER SECTION, SHOULDER ROAST, STEP 11:
I always tie a triple knot; it's very straightforward and works for me. There are many other knots that work well for roasts and portion-size cuts.

SHOULDER SECTION, SHOULDER ROAST, STEP 12:
Lightly score the skin. Because this pig is relatively small, there is some fat but not much, so shallow scoring is enough.

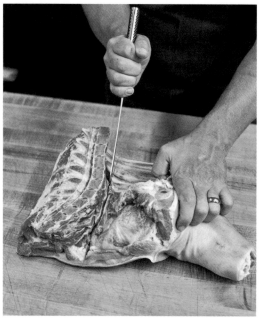

UPPER/LOWER SHOULDER SECTION, STEP 1: On the other shoulder, with your knife, lightly mark the next cut on the ribs directly below the base of the vertebrae.

UPPER/LOWER SHOULDER SECTION, STEP 2: Using the saw, cut right through the ribs.

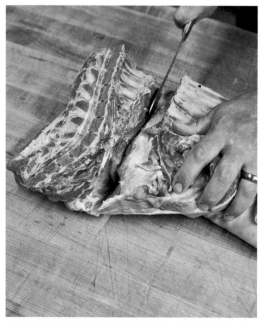

UPPER/LOWER SHOULDER SECTION, STEP 3: With the knife, finish cutting through the meat all the way to the shoulder blade.

UPPER/LOWER SHOULDER SECTION, STEP 4: Now come back with the saw and go through the shoulder blade to separate the upper and lower shoulder.

UPPER/LOWER SHOULDER SECTION, STEP 5: **Using the knife, finish cutting through the meat.**

UPPER/LOWER SHOULDER SECTION, STEP 6: **The split shoulder.**

UPPER/LOWER SHOULDER SECTION, STEP 7: **Saw at an angle to remove the chine bone.**

UPPER/LOWER SHOULDER SECTION, STEP 8: **The finished picnic ham (left), chine bone (center), and shoulder butt roast (right).**

UPPER SHOULDER SECTION, BUTT ROAST, STEP 1: Begin removing the skin at the corner of the butt roast.

UPPER SHOULDER SECTION, BUTT ROAST, STEP 2: Flip over the roast and finish cutting off the skin from the bottom. (It's easier to remove skin when working against a flat surface.)

UPPER SHOULDER SECTION, BUTT ROAST, STEP 3: **Tie at 1-inch/ 2.5-centimeter intervals right on the bone, making the first tie in the center.**

UPPER SHOULDER SECTION, BUTT ROAST, STEP 4: **On the other side of the first tie, make your third knot.**

UPPER SHOULDER SECTION, BUTT ROAST, STEP 5: **Make the fourth tie 1 inch/2.5 centimeters past the second tie. Continue adding ties, alternating sides and tying them 1 inch/2.5 centimeters beyond the last tie you placed. Alternating sides is imperative for an evenly tied roast.**

UPPER SHOULDER SECTION, BUTT ROAST, STEP 6: **In order to make equal portions, it's very important to tie the roast tightly. This affects presentation and ensures consistent cooking times.**

UPPER SHOULDER SECTION, BUTT ROAST, STEP 7: **The finished butt roast. (You could slow-roast this as is, if desired.)**

UPPER SHOULDER SECTION, BUTT ROAST, MEDALLIONS, STEP 1: **Every two ties, cut through the meat, then use a saw to cut through the bone.**

UPPER SHOULDER SECTION, BUTT ROAST, MEDALLIONS, STEP 2: The shoulder butt medallion. (You could braise, grill, or smoke this slow-and-low.)

LOWER SHOULDER SECTION, SEMI-BONELESS PICNIC HAM, STEP 1: Remove the country ribs from the picnic ham. They're great with a dry rub, then smoked until the meat falls off the bone.

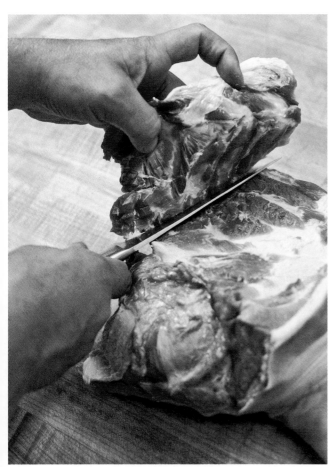

LOWER SHOULDER SECTION, SEMI-BONELESS PICNIC HAM, STEP 2:
When removing the ribs, leave a good amount of meat on the bones.

LOWER SHOULDER SECTION, SEMI-BONELESS PICNIC HAM, STEP 3:
Feel to locate the ball joint between the arm bone and the foreshank. (In this picture, my thumb is resting on the base of the shoulder blade.)

LOWER SHOULDER SECTION, SEMI-BONELESS PICNIC HAM, STEP 4:
Now make an incision, starting at the ball joint, cutting all the way
down to the shoulder blade.

LOWER SHOULDER SECTION, SEMI-BONELESS PICNIC HAM, STEP 5:
Cut downward to begin freeing the blade from the shoulder.

LOWER SHOULDER SECTION, SEMI-BONELESS PICNIC HAM, STEP 6:
Remove the shoulder blade bone at the ball joint.

LOWER SHOULDER SECTION, SEMI-BONELESS PICNIC HAM, STEP 7:
The top of the shoulder blade bone, removed from the arm bone.

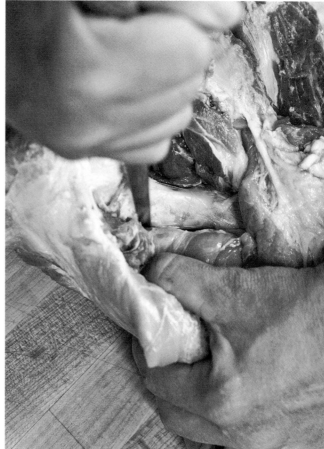

LOWER SHOULDER SECTION, SEMI-BONELESS PICNIC HAM, STEP 8:
Begin removing the arm bone from the foreshank bone.

LOWER SHOULDER SECTION, SEMI-BONELESS PICNIC HAM, STEP 9:
Keeping the knife close to the bone, remove as much meat as
possible, so that the bone will come out nice and clean.

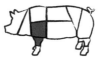

LOWER SHOULDER SECTION, SEMI-BONELESS PICNIC HAM,
STEP 10: **Now that the top of the arm bone has been freed up, work the knife down to remove the meat from the bottom of the arm bone.**

LOWER SHOULDER SECTION, SEMI-BONELESS PICNIC HAM,
STEP 11: **Make a couple more cuts to free up the arm bone from the foreshank.**

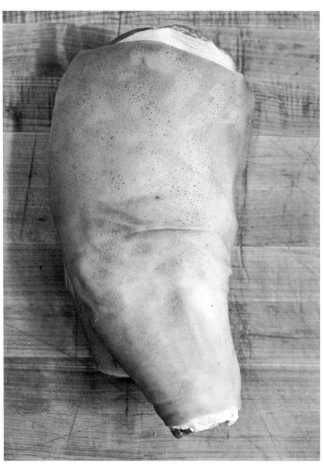

LOWER SHOULDER SECTION, SEMI-BONELESS PICNIC HAM,
STEP 12: **The finished semi-boneless, skin-on picnic ham.**

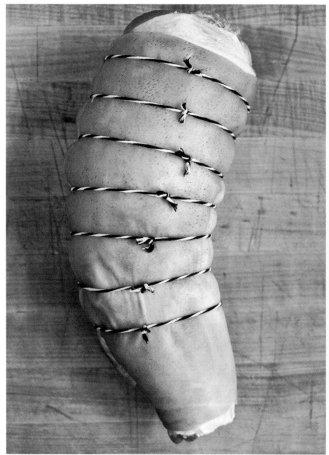

LOWER SHOULDER SECTION, SEMI-BONELESS PICNIC HAM,
STEP 13: **Picnic ham, tied to keep it compact. (For cooking, you would season it before tying.)**

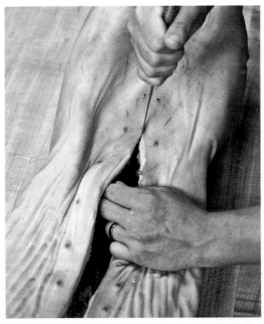

LOIN/SADDLE SECTION, STEP 1: Cut through the belly, continuing the incision all the way down to the hams. This will allow you to see into the cavity and plan your next cut.

LOIN/SADDLE SECTION, STEP 2: The first incision will be made between the 13th and 14th ribs (second to last, and last).

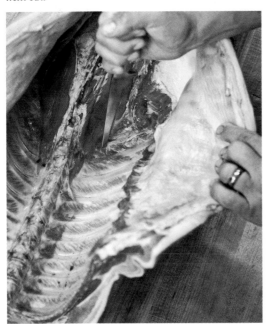

LOIN/SADDLE SECTION, STEP 3: Beginning the cut that will separate the loin and the belly from the saddle, cut between the 13th and 14th ribs.

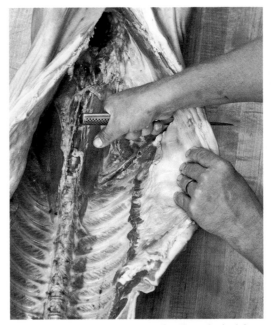

LOIN/SADDLE SECTION, STEP 4: Continue the incision to the edge of the belly, following the angle of the ribs.

LOIN/SADDLE SECTION, STEP 5: Finish the incision, cutting through the meat only.

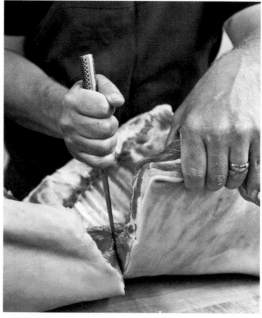

LOIN/SADDLE SECTION, STEP 6: Repeat the process on the other side, using the knife to cut through the meat only, not the bones.

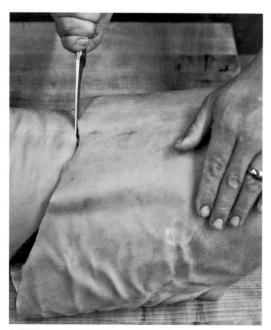

LOIN/SADDLE SECTION, STEP 7: Flip over the loin saddle. With the knife, connect the cuts on the top side of the animal.

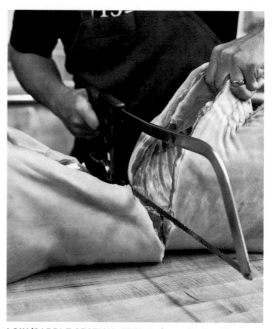

LOIN/SADDLE SECTION, STEP 8: Come back with a saw to cut through the spine, gently pulling apart the saddle and belly to ease the saw's progress.

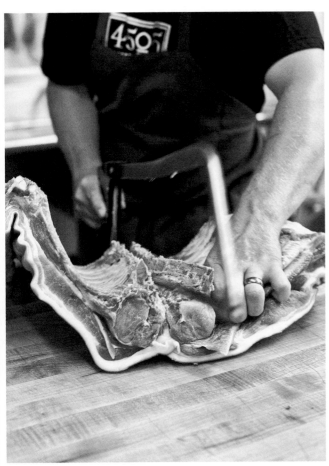

LOIN/BELLY SECTION, STEP 1: Saw straight down the center of the spine to make two equal sides.

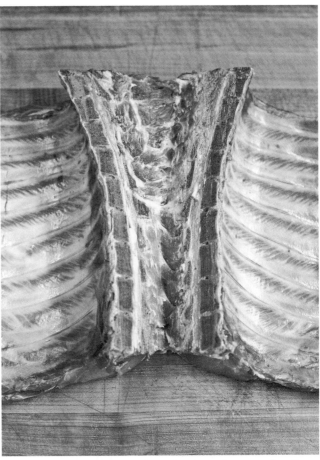

LOIN/BELLY SECTION, STEP 2: Note that the cut has gone directly through the bone only.

LOIN/BELLY SECTION, STEP 3: **Finish separating the two sides, using a knife to cut through the remaining flesh.**

LOIN/BELLY SECTION, STEP 4: **Measure the eye of the meat on one side of the loin section.**

LOIN/BELLY SECTION, STEP 5: **Add half of that distance to determine where to mark the end of your cut.**

LOIN/BELLY SECTION, STEP 6: **Make the mark to guide you when splitting apart the loin and the belly section. This will help you avoid cutting into the loin.**

LOIN/BELLY SECTION, STEP 7: This is the triangle of goodness, a little piece of dark meat that collects a lot of the tasty caramelization from the hot grill or pan-searing—my favorite part of the pork chop.

LOIN/BELLY SECTION, STEP 8: Measure and mark the other rack, as in steps 4 to 6.

LOIN/BELLY SECTION, STEP 9: Connect the marks on the back of the rib rack.

LOIN/BELLY SECTION, STEP 10: Continue cutting with a saw, to get through the rib bones.

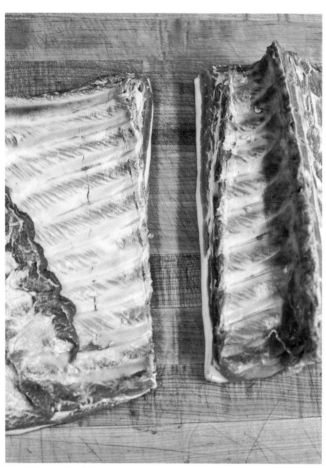

LOIN/BELLY SECTION, STEP 11: Here, the belly has been separated from the bone-in pork rack.

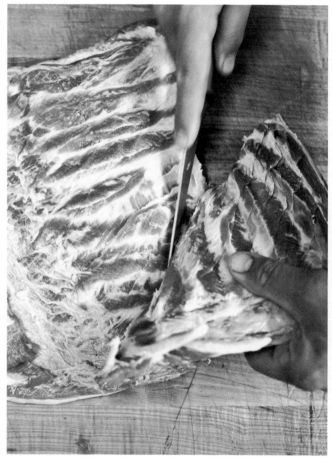

LOIN/BELLY SECTION, STEP 12: Keeping a good amount of meat on the ribs (about ¼ inch/6 millimeters of meat), peel back and remove all the ribs from the belly.

LOIN/BELLY SECTION, STEP 13: **From the top: The separated bone-in rack, the rack of ribs (from which we've removed the cartilage that you see right below), and the boneless belly.**

LOIN SECTION, BONE-IN SKIN-ON CHOPS, STEP 1: On the bone-in rack of pork, use a saw to cut between each rib through the chine bone, cutting through the bone only, not the flesh.

LOIN SECTION, BONE-IN SKIN-ON CHOPS, STEP 2: With the knife, finish cutting down through the flesh and skin to release each chop.

LOIN SECTION, BONE-IN SKIN-ON CHOPS, STEP 3: The finished bone-in skin-on pork rib chops. I like to keep the skin on and grill the chop, finishing them off with the skin-side down, over red-hot coals. Crispy skin is always the way to go.

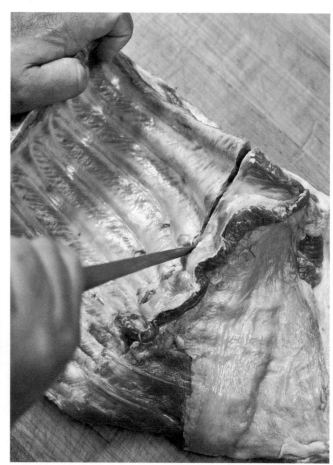

LOIN/BELLY SECTION, FRENCHED PORK RACK, STEP 1: **This is where the tip of the rib is located. A little bulge indicates where the cartilage begins.**

LOIN/BELLY SECTION, FRENCHED PORK RACK, STEP 2: **Make an incision with the knife, cutting through the cartilage, to separate the cartilage from the rib.**

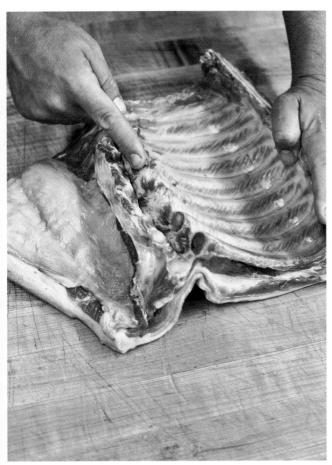

LOIN/BELLY SECTION, FRENCHED PORK RACK, STEP 3: Here is the cartilage; be sure to cut through only the cartilage—not all the way through the meat of the belly—as I did for the first two ribs.

LOIN/BELLY SECTION, FRENCHED PORK RACK, STEP 4: With the knife, remove all the white cartilage from the tip of each bone, so that the inside of the bone is exposed and dry looking.

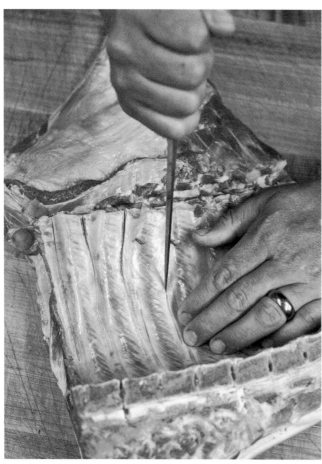

LOIN/BELLY SECTION, FRENCHED PORK RACK, STEP 5: Carefully press down hard with the knife, making an incision all along the back of each bone and cutting through the sinew right in the center of each bone.

LOIN/BELLY SECTION, FRENCHED PORK RACK, STEP 6: Peeling back the sinew and meat from the top and back of each rib will make the frenching process much easier.

LOIN/BELLY SECTION, FRENCHED PORK RACK, STEP 7: It's very important that each rib tip be completely dry and clean and free of all meat and sinew; you must complete the process on each rib, in order, before you move on to the next one.

LOIN/BELLY SECTION, FRENCHED PORK RACK, STEP 8: Use one hand to brace against the loin, without pushing down too hard, and form a "C" with your finger on the tip of the rib.

LOIN/BELLY SECTION, FRENCHED PORK RACK, STEP 9: Pull down toward the loin with that "C," and the rib should pop right out. (Unless you follow this process exactly, you'll be scraping the sinew off the ribs all day.)

LOIN/BELLY SECTION, FRENCHED PORK RACK, STEP 10: Here, the ribs have been popped out, leaving all the rib meat cleanly behind.

LOIN/BELLY SECTION, FRENCHED PORK RACK, STEP 11: **Before cutting, make your marks as we did on pages 184–185. Connect the marks by cutting straight through the belly.**

LOIN/BELLY SECTION, FRENCHED PORK RACK, STEP 12: **The frenched rack, separated from the belly and rib meat.**

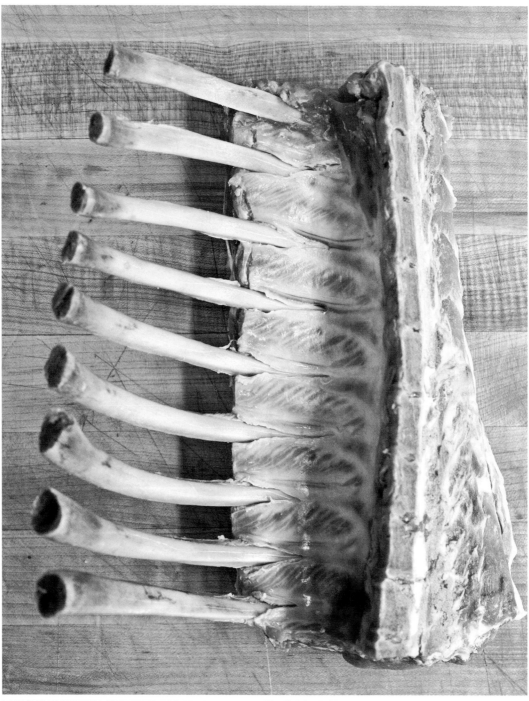

LOIN/BELLY SECTION, FRENCHED PORK RACK, STEP 13: The finished skin-on frenched pork rack with the chine intact. Roasting with the meat side uppermost allows the tips of the ribs and chine bone to act as a rack, protecting the delicate meat.

LOIN SECTION, FRENCHED PORK CHOPS, STEP 1: Remove the skin from the rack, starting at the corner of the loin.

LOIN SECTION, FRENCHED PORK CHOPS, STEP 2: Come back with the saw to cut down through the chine bone between every rib first (not through the meat).

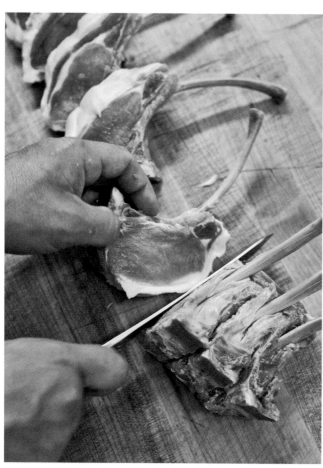

LOIN SECTION, FRENCHED PORK CHOPS, STEP 3: Come back with the knife to finish cutting the flesh, freeing up individual chops.

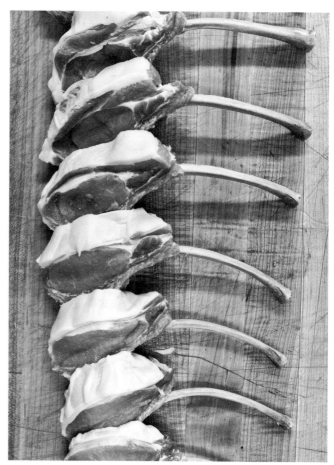

LOIN SECTION, FRENCHED PORK CHOPS, STEP 4: The finished frenched pork rib chops—beautiful.

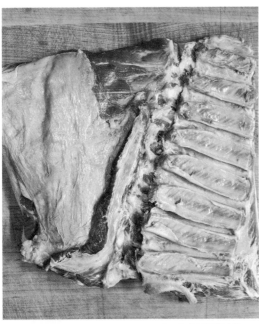

BELLY SECTION, BELLY STRIPS WITH RIB MEAT ON, STEP 1: The pork belly with rib meat. Most people don't save the rib meat when frenching, but it's dee-licious!

BELLY SECTION, BELLY STRIPS WITH RIB MEAT ON, STEP 2: Back to the belly and rib meat. Remove the rib cartilage that was left behind in the frenching process.

BELLY SECTION, BELLY STRIPS WITH RIB MEAT ON, STEP 3: Skin-on pork belly with rib meat. Tip: I never take off that rib meat, unless I am making bone-in ribs.

BELLY SECTION, BELLY STRIPS WITH RIB MEAT ON, STEP 4: Using the raised divisions of the rib meat as guides, cut the individual belly strips.

BELLY SECTION, BELLY STRIPS WITH RIB MEAT ON,
STEP 5: **The boneless belly strips.**

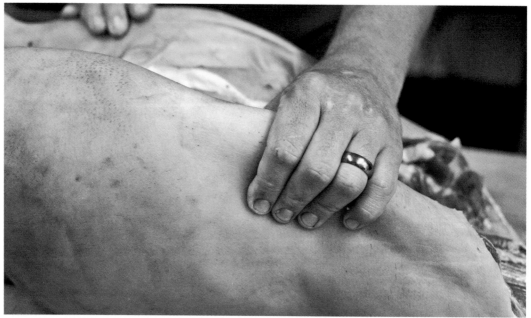

SADDLE/LEG SECTION, STEP 1: Pinch the skin to locate where the leg muscle begins and ends. This is where
you will make the mark for the first cut to separate the leg and saddle.

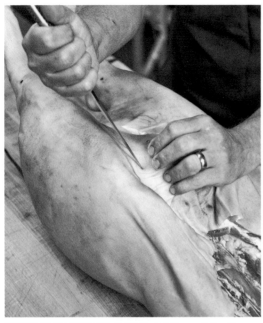

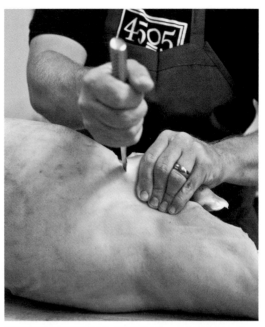

SADDLE/LEG SECTION, STEP 2: Cut through the flap section toward the belly to release the saddle belly from the leg.

SADDLE/LEG SECTION, STEP 3: The next steps can be tricky, so look ahead to page 204 to see the end result before you continue. Cutting the skin and meat, cut along the edge of the leg.

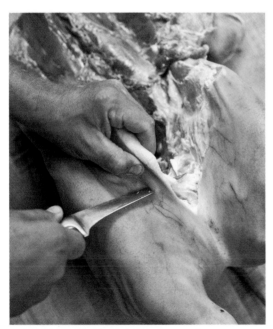

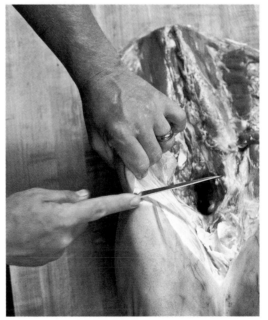

SADDLE/LEG SECTION, STEP 4: Repeat the process on the other side. Come in with the knife and make the starter cut away from the direction you want to go.

SADDLE/LEG SECTION, STEP 5: Come back and cut down along the inside of the leg, just as before.

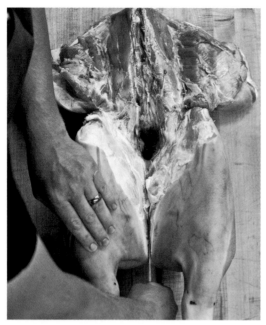

SADDLE/LEG SECTION, STEP 6: The entire saddle belly has been removed from the leg section, so now it's time to concentrate on the tenderloins.

SADDLE/LEG SECTION, STEP 7: Follow the length of the tenderloin all the way up to the inside of the leg. Come in with the knife by the leg and make a small incision to free up the butt end of the tenderloin.

SADDLE/LEG SECTION, STEP 8: Fillet the tenderloin back far enough to avoid sawing through it when we remove the saddle section. (Look ahead to Sadde Section, Step 1; you can see the tenderloins extended on the cutting board.)

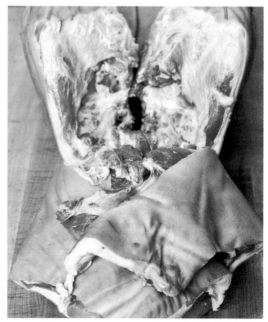

SADDLE/LEG SECTION, STEP 9: The tenderloin has been freed up from the pelvic bone, leaving the pelvic bone exposed, so that you can saw through it.

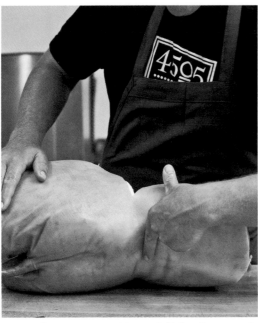

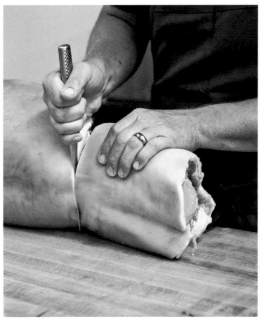

SADDLE/LEG SECTION, STEP 10: Roll the pig over and feel the top of the pelvic bone. This is where you'll make the incision.

SADDLE/LEG SECTION, STEP 11: Cut through the flesh all the way to the top of the pelvic bone.

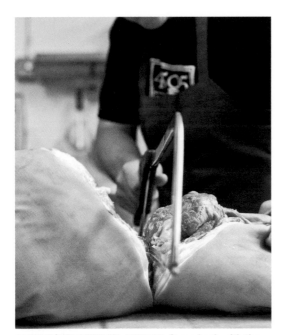

SADDLE/LEG SECTION, STEP 12: Come back with the saw to cut through the pelvic bone.

SADDLE SECTION, STEP 1: Now that the saddle has been removed from the leg, go ahead and saw straight through the center, leaving two equal halves.

SADDLE SECTION, STEP 2: **Come back with the knife to finish up cutting through the meat.**

SADDLE SECTION, STEP 3: **The two halves of the saddle.**

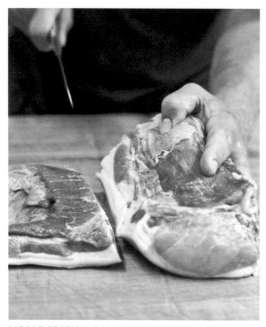

SADDLE SECTION, PORTERHOUSE, STEP 1: **Remove and square off the saddle belly.**

SADDLE SECTION, PORTERHOUSE, STEP 2: **Cut the top of the saddle all the way to the bone to portion out the porterhouses.**

SADDLE SECTION, PORTERHOUSE, STEP 3: Saw through the backbones only, not the flesh.

SADDLE SECTION, PORTERHOUSE, STEP 4: The sawing of the porterhouse section has been done.

SADDLE SECTION, PORTERHOUSE, STEP 5: Finish the cut with the knife, cutting through the rest of the meat.

SADDLE SECTION, PORTERHOUSE, STEP 6: Remove the first steak with a nice, clean cut.

SADDLE SECTION, PORTERHOUSE, STEP 7: The finished, skin-on porterhouse and T-bone steaks.

SADDLE SECTION, TENDERLOIN, STEP 1: On the other side of the saddle section, take your knife and cut along the backbone without cutting through the tenderloin, slowly peeling it off.

SADDLE SECTION, TENDERLOIN, STEP 2: Peel the tenderloin away from the bone, leaving as much meat as possible on the loin.

SADDLE SECTION, TENDERLOIN, STEP 3: The pork tenderloin. I don't like to trim it down any further than this. More fat = more flavor.

SADDLE SECTION, TENDERLOIN, STEP 4: Cut the tenderloin medallions the width of two fingers (about 2 inches/5 centimeters).

SADDLE SECTION, TENDERLOIN, STEP 5: The finished tenderloin medallions.

SADDLE SECTION, BONELESS LOIN, STEP 1: The 14th rib (the floating rib) is still on the saddle section, and it's in the way; so, remove it.

SADDLE SECTION, BONELESS LOIN, STEP 2: With a rolling motion, cut away all the bone from the loin.

SADDLE SECTION, BONELESS LOIN, STEP 3: Continue to remove the bones, leaving as much meat as possible on the loin.

SADDLE SECTION, BONELESS LOIN, STEP 4: The backbone has been fully removed from the saddle.

SADDLE SECTION, BONELESS LOIN, STEP 5: Place the loin skin-side down, and start cutting at the corner to fillet the meat from the skin.

SADDLE SECTION, BONELESS LOIN, STEP 6: Peel the loin back bit by bit, cutting as you go, leaving as much fat as possible on the loin and as little as possible on the skin.

SADDLE SECTION, BONELESS LOIN, STEP 7: **The finished boneless pork loin.**

SADDLE SECTION, BONELESS LOIN CHOPS, STEP 1: **Measure the width of 2 fingers (about 2 inches/5 centi-meters), and holding the meat firmly, make a smooth, straight cut with your knife.**

SADDLE SECTION, BONELESS LOIN CHOPS, STEP 2: **Portioning the boneless pork loin chops.**

SADDLE SECTION, BONELESS LOIN CHOPS, STEP 3: **The finished boneless pork loin chops.**

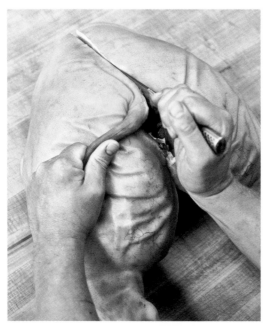

LEG SECTION, STEP 1: **To split the leg section in half, make the first cut to one side of the tail with your knife.**

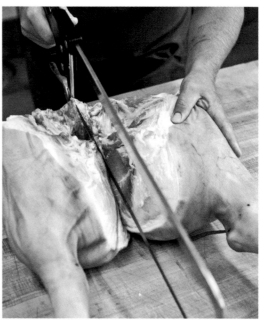

LEG SECTION, STEP 2: **Come back with the saw to cut through the bone.**

LEG SECTION, STEP 3: **Carefully brace the legs, so that the meat doesn't tear when you separate the two sides.**

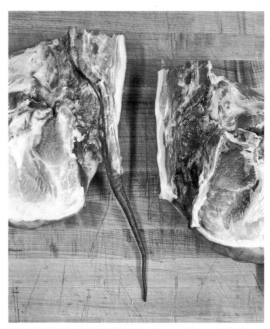

LEG SECTION, STEP 4: **The two leg sections, with the tail on one side.**

LEG SECTION, TROTTER, STEP 1: Remove the trotter by cutting above the hock joint; work the joint back and forth to find the soft spot.

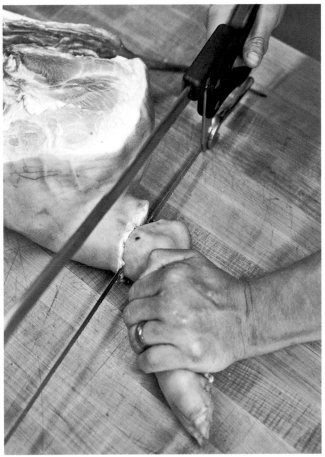

LEG SECTION, TROTTER, STEP 2: Come back with the saw and cut through the bone.

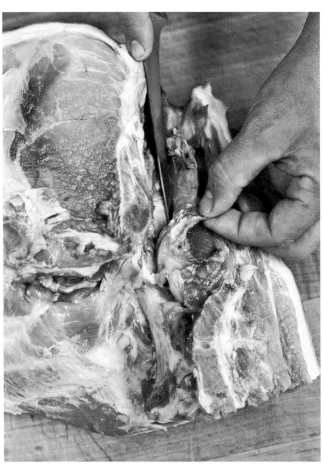

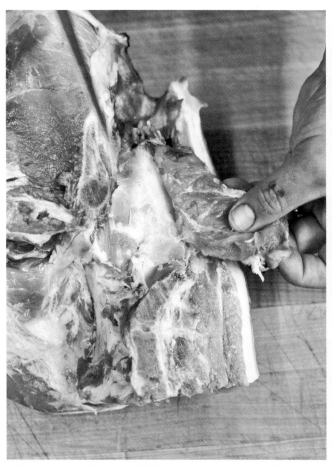

LEG SECTION, SPIDER CUT, STEP 1: Cut on the outside of the pelvic bone to release the little piece of meat between the aitch bone and the hip bone.

LEG SECTION, SPIDER CUT, STEP 2: Pull away this cut, called the spider cut.

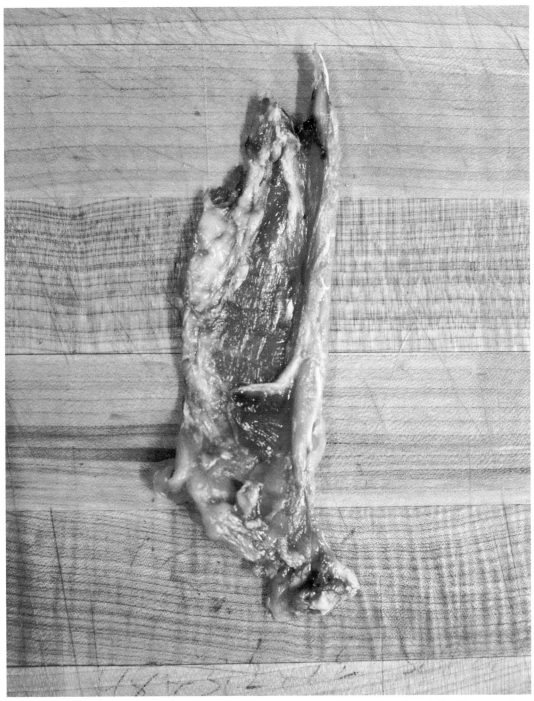

LEG SECTION, SPIDER CUT, STEP 3: The spider cut is usually left behind or goes into the grinder; it's exposed to the air, so it may dry out, but when you're butchering your own animal, it makes a great snack.

LEG SECTION, SEMI-BONELESS HAM, STEP 1: **This is another tricky sequence. Look ahead at the next six steps to see what you'll be doing. Cut behind the hip bone to free up the meat, leaving as much meat as possible on the leg and keeping the bone nice and clean.**

LEG SECTION, SEMI-BONELESS HAM, STEP 2: **Work the knife into the ball socket to cut through the ligament between the ball socket and the ball joint.**

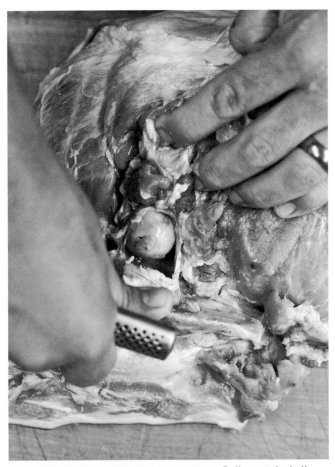

LEG SECTION, SEMI-BONELESS HAM, STEP 3: Pull apart the ball joint to begin freeing the pelvic bone. This will help you see and plan your next cut.

LEG SECTION, SEMI-BONELESS HAM, STEP 4: Come back with your knife to finish removing the pelvic bone, keeping one side of the knife on the bone at all times.

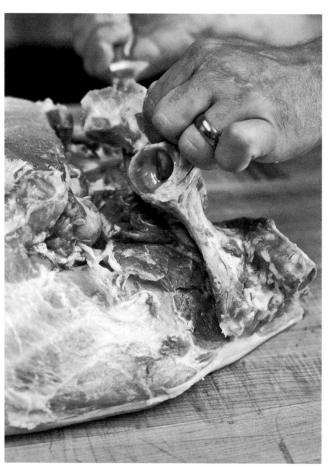

LEG SECTION, SEMI-BONELESS HAM, STEP 5: Work the bone back and forth, and remove it as cleanly as possible to maximize the yield of usable sirloin and leg meat.

LEG SECTION, SEMI-BONELESS HAM, STEP 6: Fully removed pelvic bone—finish up the last cut to release it.

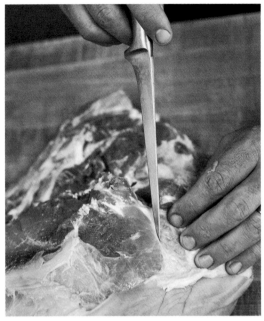

LEG SECTION, SEMI-BONELESS HAM, STEP 7: **Map out the next cut before beginning, locating the point between the leg muscles where you will begin the incision.**

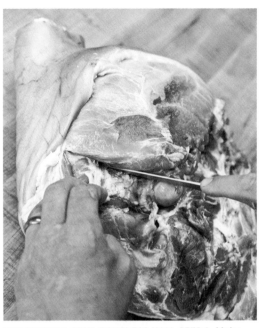

LEG SECTION, SEMI-BONELESS HAM, STEP 8: **Make the first cut straight down between the muscles to the bone.**

LEG SECTION, SEMI-BONELESS HAM, STEP 9: **Free the leg bone from the flesh.**

LEG SECTION, SEMI-BONELESS HAM, STEP 10: **Continue cutting and moving the knife around to free up the leg bone from the hind shank bone.**

LEG SECTION, SEMI-BONELESS HAM, STEP 11: **Make the final cut to remove the bone.**

LEG SECTION, SEMI-BONELESS HAM, STEP 12: **Score the skin with very shallow cuts. I prefer to cut in one direction to keep the skin tight and prevent it from shrinking into tiny squares.**

LEG SECTION, SEMI-BONELESS HAM, STEP 13: **Pull butcher's twine through a large beef-trussing needle.**

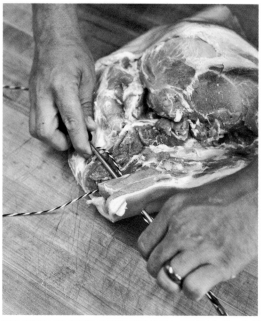

LEG SECTION, SEMI-BONELESS HAM, STEP 14: **Carefully push the trussing needle through the skin, starting at the bottom of the ham. Truss up the bottom of the ham only (not all around); this will help the roast cook evenly. (The needle is penetrating the sirloin here; I always keep the sirloin on for this wonderful roast.)**

LEG SECTION, SEMI-BONELESS HAM, STEP 15: **Leave plenty of room in between each puncture point; this will help pull the skin evenly, all around the ham.**

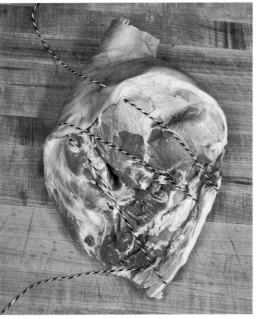

LEG SECTION, SEMI-BONELESS HAM, STEP 16: **The semi-boneless ham—the only bone still in the roast is the hind shank bone. The top of this bone will pop out during cooking.**

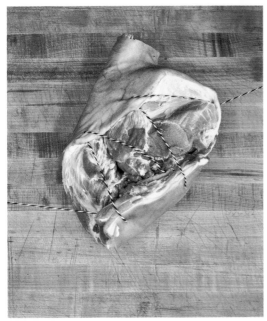

LEG SECTION, SEMI-BONELESS HAM, STEP 17: **Before tying off, pull the trussing string nice and tight, and tie with a triple knot in the center.**

LEG SECTION, SEMI-BONELESS HAM, STEP 18: **The scored, semi-boneless ham, with the sirloin on.**

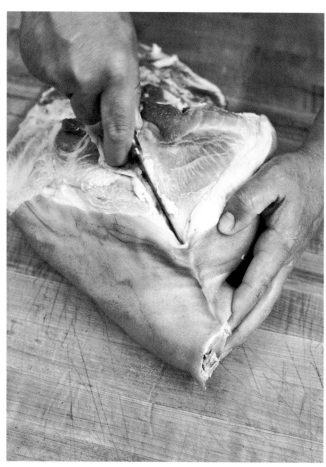

LEG SECTION, SIRLOIN, STEP 1: On the other side of the leg, remove the trotter as before, then cut right above the shank, with the blade up, to begin removing the skin.

LEG SECTION, SIRLOIN, STEP 2: Slowly ease the skin off the ham, peeling and making little shallow cuts; leave as much fat as possible on the ham.

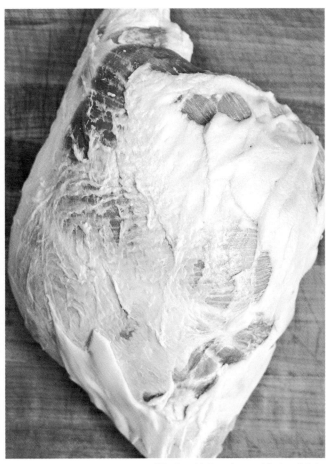

LEG SECTION, SIRLOIN, STEP 3: Whole, skinless ham with the sirloin on and the shank on.

LEG SECTION, SIRLOIN, STEP 4: Right below the leg bone joint, make the first cut straight down to remove the sirloin.

LEG SECTION, SIRLOIN, STEP 5: Continue the cut down from the top all the way through the sirloin.

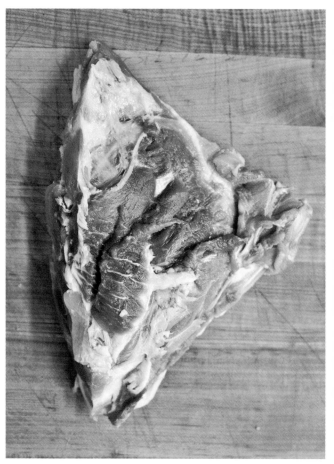

LEG SECTION, SIRLOIN, STEP 6: The whole pork sirloin is so tasty when brined and smoked (see page 236). A smoky treat! It's an excellent—and leaner—substitute for bacon.

LEG SECTION, COWBOY STEAKS, STEP 1: **Divide the leg into three 2-inch-/5 centimeter-thick steaks: First, cut down to the bone only. (Cuting around on both sides now would make the piece too floppy to saw.)**

LEG SECTION, COWBOY STEAKS, STEP 2: **Make the second cut, down from the top.**

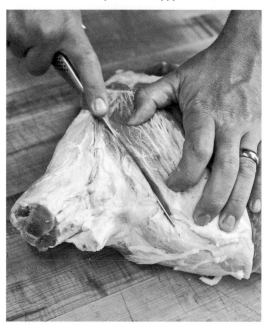

LEG SECTION, COWBOY STEAKS, STEP 3: **Make the last cut, creating the third steak; this cut should come right above the kneecap, where the hind shank and the leg bone meet.**

LEG SECTION, COWBOY STEAKS, STEP 4: **All the cuts have been made down to the bone only, and you're ready to saw.**

LEG SECTION, COWBOY STEAKS, STEP 5: Come back and saw through the leg bone.

LEG SECTION, COWBOY STEAKS, STEP 6: Now come back with the knife and make nice, clean cuts through the meat to separate the three leg (cowboy) steaks.

LEG SECTION, COWBOY STEAKS, STEP 7: Make the cut between the shank of the leg joint. If you're in the middle, you should be able to cut right through.

LEG SECTION, COWBOY STEAKS, STEP 8: The three finished cowboy steaks, always spectacular with a dry rub and cooked slow-and-low.

Charred Scallion Sausage

YIELD: 4.3 pounds/2 kilograms

These tasty sausages are one of my very favorite uses for pig skin. The skin provides amazing body and richness, keeping the meat juicy and emulsified. Because bacon fat has a low melting point, it helps maintain that incredibly juicy character. The charred scallions add a super-memorable sweet-smoky flavor—and the hint of ginger is a great accent. Char the scallions on a grill until tender and nicely caramelized. To make the sausages, you will need about 25 feet/7.5 meters of large hog casings, which can be ordered by the hank (about 100 feet/30 meters) from a specialty butcher or on the Internet (see page 237).

Large hog casings	About 25 feet/7.5 meters			
DRY INGREDIENTS				
Fine sea or kosher salt	4 tsp	0.8 oz	22 g	1.1%
Granulated sugar	½ tsp	0.1 oz	3 g	0.2%
Freshly ground black pepper	1 tbsp	0.2 oz	6 g	0.3%
Ground fennel seeds	1½ tsp	0.1 oz	3 g	0.1%
Dried red chile flakes	¼ tsp	0.1 oz	3 g	0.1%
MEAT				
Pork shoulder and/or trim, very cold	1 whole	2.5 lb	1,138 g	57.9%
Pork skin, boiled until tender (about 6 hours), and chilled until very cold	n/a	4.6 oz	131 g	6.7%
Bacon, very cold	n/a	11.2 oz	317 g	16.1%

cont'd

WET MIX				
Jalapeño, seeded and minced	1 small	0.3 oz	9 g	0.4%
Charred scallions, trimmed and minced	2 bunches	7.2 oz	205 g	10.3%
Freshly grated ginger (on a microplane)	½ tsp	0.1 oz	3 g	0.2%
Minced garlic	1 tsp	0.2 oz	7 g	0.4%
Finely chopped cilantro	5 sprigs	0.4 oz	11 g	0.6%
Ice water	½ cup	3.9 oz	109 g	5.6%

1. Read the General Sausage Making Tips (see page 17).

2. The night before: Soak the hog casings in a bowl of cold water; refrigerate overnight.

3. Assemble all of the dry ingredients in a container. (This step need not be done the night before, but it's crucial that it be completed before you start grinding the meat.)

4. The next day: Untangle the casings and begin to open them to make the stuffing process easier. Hold one end of each piece of casing up to the nozzle of the faucet and support it with your other hand. Gently turn on the water and let it run through the casings to check for holes. If there are any holes in the casings, cut out the pieces with the holes. Hold the casings in a bowl of ice water or refrigerate until stuffing time.

5. With a sharp boning knife or your knife of choice, remove the meat and fat from the bones, if necessary. Open-freeze the meat, uncovered, for 30 to 60 minutes, until the surface of the meat is crunchy to the touch and the interior is very cold, but not frozen.

6. Cut all the pork, pork skin, bacon, and trotter into 1-inch/2.5-centimeter-square cubes or a size slightly smaller than the opening of the meat grinder. Open-freeze the meat again, uncovered, for 30 to 60 minutes, until the surface of the meat is crunchy to the touch and the interior is very cold, but not frozen.

7. When you are ready to grind, prepare a perfectly clean and chilled meat grinder for grinding, and fit it with the large plate. Grind the pork first (leaving the pork skin and bacon in the freezer). Start the auger and, without using the supplied pusher, let the auger gently grab each cube of meat and bring it forward toward the blade and through the grinding plate. Continue grinding until all of the meat has been processed. Place it in a clean, cold nonreactive bowl or tub and again open-freeze, uncovered, for 30 to 60 minutes, until the surface of the meat is crunchy to the touch and the interior is very cold, but not frozen. Switch to the small grinding plate and grind the skin and bacon. Add to the open-freezing pork.

8. In a medium nonreactive bowl, combine the dry ingredients and wet mix, and whisk together until completely blended and the dry ingredients have dissolved. I call this the "slurry."

9. In a large, wide basin or bowl that will give you plenty of room to mix the meat and seasonings, combine the cold meat with the slurry. Roll up your sleeves and, with perfectly clean hands, begin kneading and turning the mixture as you would a large quantity of bread dough. Eventually, you will begin to notice that the mixture has acquired a somewhat creamy texture. This is caused by the warmth of your hands and is a sign that you have finished mixing. Spoon out a few tablespoons of the mixture, and return the remainder to the refrigerator.

10. In a nonstick skillet over medium heat, lightly fry a test portion of sausage mixture until cooked through but not caramelized (which would change the flavor profile). Taste for seasoning. Based on this taste test, you can adjust the salt in the main portion of sausage, if desired.

11. Prepare a perfectly clean and chilled sausage stuffer and place the water-filled bowl of casings next to it. You will also need a landing surface of clean trays or parchment paper–lined baking sheets for your finished sausages.

12. Load the sausage mixture into the canister of the sausage stuffer, compacting it very lightly with a spatula to be sure there are no air pockets. Replace the lid.

13. Thread a length of casing all the way onto the stuffing horn and start cranking just enough to move a little of the ground meat mixture into the casing. As soon as you can see the meat poking through the nose of the stuffer, stop and crank backward slightly to halt the forward movement. Pinch the casing where the meat starts (to extrude all the air), and tie into a knot. Now start cranking again with one hand, while you support the emerging sausage with the other. Move the casing out slowly to allow it to fill fully but not too tightly, so that there will be some give in the sausage when it comes time to tie the links. When you get close to the end, leave 6 inches/15 centimeters of unstuffed casing and stop cranking.

14. Go back to the original knot and measure 6 inches/15 centimeters of sausage. Pinch the sausage gently to form your first link, and twist forward for about seven rotations. Move another 6 inches/15 centimeters down the sausage, and this time, pinch firmly and twist backward. Repeat this process every 6 inches/15 centimeters, alternating forward and backward, until you reach the open end of the casing. Twist the open end right at the last bit of sausage to seal off the whole coil, and then tie a knot.

15. Ideally, hang the sausage overnight in a refrigerator, or refrigerate on parchment paper–lined baking sheets, covered with plastic wrap, to allow the casing to form fully to the meat, and the sausage to settle. (Or, if desired, you can cook the sausages right away.) The next day, cut between each link and cook as desired.

Pork Jowl and Clams

SERVES 5

Pork jowl and clams is definitely one of my favorite flavor combos.
It was also a favorite of my late dear friend, talented chef Devin Autrey.
This dish always brings back such good memories of when we cooked
together—thank you, Devin. The addition of white wine and clams
cuts through the tasty, rich fat of the jowl, and with the crunchy skin—
wow!—talk about Flavor Town. You can also substitute pork belly for
the jowl and serve it with *chicharrones* (crispy pork rinds) on top. Enjoy!

JOWL				
Pork jowl (see page 156), skin-on and trimmed of excess fat	1 whole	1.8 lb	826 g	29.8%
Fine sea salt	3½ tbsp	2.1 oz	60 g	2.2%
Granulated sugar	2 tbsp	1 oz	27 g	1%
Coarsely ground black pepper, plus extra for seasoning	1 tbsp	0.3 oz	8 g	0.3%
Ground coriander seeds	1 tbsp	0.1 oz	4 g	0.1%
Extra-virgin olive oil	as needed			
CLAMS				
Live Manila clams	3 lb	48 oz	1, 362 g	49.1%
Thickly sliced shallots	1 cup	4.6 oz	130 g	4.7%
Fresh thyme sprigs, tied together	10	0.2 oz	7 g	0.3%
Garlic, sliced as thick as a nickel	3 large cloves	0.7 oz	20 g	0.7%
Fine sea salt	1 tsp	0.3 oz	5 g	0.1%
White wine	¾ cup + 2 tbsp	7.1 oz	200 g	7.2%
Dried red chile flakes	¾ tsp	0.1 oz	2 g	0.1%

Unsalted butter, very cold, cut into small pieces	½ cup	4 oz	112 g	4%
Chiffonade of flat-leaf parsley	¼ cup	0.2 oz	7 g	0.3%
Fresh lemon juice (preferably Meyer) Lemon wedges (for serving)	1 tsp	0.1 oz	4 g	0.1%

1. Score the flesh side (not the skin) of the jowl in 1-inch/2.5-centimeter squares, cutting 1 inch/2.5 centimeters deep. In a small bowl, combine the salt, sugar, pepper, and coriander seeds. Rub about one-third of the seasoning mixture into the fat side of the jowl. Place, fat-side down, on a rack over a tray or sheet pan and sprinkle with the remaining seasoning mixture. Massage it evenly into the scored cuts, packing it in—and on—firmly. Refrigerate the jowl for 6 hours.

2. Preheat the oven to 450°F/230°C/gas 8. Rinse the jowl lightly under cold running water, leaving a little of the curing mixture on the meat, and pat dry thoroughly with paper towels. Paint the skin with olive oil and place skin-side up on a rack over a roasting pan. Roast for 10 minutes. Turn down the oven to 225°F/110°C/gas ¼ and slow-roast for 30 minutes more, or until the internal temperature reaches 180°F/82°C. Once the skin is crisp, remove the jowl from the oven and allow it to rest for 30 to 60 minutes. Cut the skin off the jowl, and reserve the skin until serving time. Cut the meat into spoon-size lardons.

3. Purge the clams by rinsing them under cold running water for about 10 minutes; drain.

4. Season the lardons generously with coarse black pepper and sizzle them in a large pot over medium heat until they begin to crisp and render a fair amount of fat. Using a slotted spoon, transfer the crispy lardons to paper towels (leaving the fat in the pot). Reduce the heat to low and add the shallots, thyme, and garlic. When the shallots are slightly softened, add the clams and shake the pot to coat them nicely with all the flavorings. Add the salt and wine and bring to a *very slow* simmer. Cook very gently—do not boil—until the clams begin to open. As they open, transfer them to a big platter with tongs and keep cooking until all of the clams have opened. (Discard any that have not opened.)

5. Add the chile flakes and butter to the pot. Swirl the pot and bring the broth to a simmer. Return the clams and lardons to the pot. Add the parsley and lemon juice, and gently fold everything together so all the ingredients get melded.

6. Transfer everything to the platter and pour all the buttery juices over the top. Break the crispy skin into small pieces and scatter them over the top. Serve with lemon wedges on the side and enjoy.

Pork Belly and Garbanzo Soup

SERVES 6

My nickname for this dish is Banzos and Bellies in Broth.
The cut used here is pork belly with rib meat on. I love to eat this
soupy dish with *chicharrones* and a big glass of pilsner. Keeping
the skin on really adds to the body of the soup, but it's not always
easy to get and the dish is great without it, too.

PORK BELLY				
Master Brine (page 16), completely cold	10 cups	80.1 oz	2,270 g	55.6%
Boneless pork belly with rib meat on (see page 202), cut into 1-inch/ 2.5 centimeter strips	1.2 lb	18.5 oz	525 g	12.9%
BEANS				
Dried garbanzo beans	1 lb	16 oz	454 g	11.1%
Water	as needed			
Carrots, peeled	2 large	8.5 oz	240 g	5.9%
Onion, halved through the root end, roots trimmed off	1 large	12.9 oz	365 g	8.9%
Garlic, loose papery skin removed and top one-third trimmed off	1 whole head	1.9 oz	55 g	1.3%
Fresh thyme sprigs, tied together	1 small bunch	1.2 oz	35 g	0.9%
Fine sea salt	2 tsp	0.4 oz	12 g	0.3%
ANCHOVY PESTO				
Oil- or salt-packed anchovies, drained	20 fillets	2.1 oz	60 g	1.5%
Fresh lemon juice	3 tbsp	1.4 oz	40 g	1%
Roughly chopped flat-leaf parsley	2 tbsp	0.4 oz	10 g	0.2%
Chile oil	1¼ tbsp	0.6 oz	18 g	0.4%

1. In a deep nonreactive container that will fit in your refrigerator, pour the cold brine over the belly strips. Place a piece of parchment paper over the top, so that no part of the meat will be exposed to oxygen. Weight with a clean plate, if necessary, to keep the meat completely submerged. Cure in the refrigerator for 20 hours.

2. At the same time, soak the garbanzo beans in water to cover generously—about three times their volume—for at least 12 hours, or overnight, and then drain.

3. Wipe off the belly strips, removing the peppercorns from the brine. In a pot, combine the drained beans, pork belly strips, carrots, onion, garlic, and thyme. Add enough water to cover by 3 inches/7.5 centimeters and bring to a boil over high heat. Reduce the heat so that the liquid barely simmers and skim off the foam. Simmer *very* gently (the liquid should just barely quiver), partially covered, for 2 to 2½ hours. When the beans reach the al dente stage, stir in the salt and simmer until tender. Let the belly broth and beans cool to room temperature all together.

4. Spoon out the carrots, onion, belly strips, and head of garlic. Cut the carrots into thick rounds; if they're too big to fit on a spoon, cut them crosswise into quarters. Cut the onion into bite-size pieces. Slice the belly strips in half lengthwise and then again crosswise into spoon-size lardons. Return the carrots, lardons, and onions to the bean broth and squeeze in the soft cloves of garlic loveliness. Discard the bare thyme sprigs.

5. To make the anchovy pesto: in a mortar and pestle, combine the anchovies, half of the lemon juice, the parsley, and chile oil and pound to a paste. (Or finely chop the anchovies and parsley and mix them with the lemon juice and oil.)

6. Warm the brothy beans and stir in the anchovy pesto. Add a bit of the remaining lemon juice to bring a little sunshine to the dish. Taste for seasoning. Enjoy.

Crispy Pork Shoulder with Shank

SERVES 10

I love this tender shoulder for juicy pork sandwiches; adding some pickled
jalapeño, cilantro, and basil, and putting it all on a warm crunchy roll
makes it taste great at any time of day. This is a really basic dry rub
that's great for slow-roasting any protein, but it's especially spectacular
with pork shoulder. Keeping the skin on gives us the crispy, crunchy
texture we all crave—it gives me goose bumps just thinking about it!

Pork picnic shoulder (see page 180), shank on and semi-boneless	1 whole	8.5 lb	3,950 g	92%
Fine sea salt (for seasoning rub)	1½ tbsp	1 oz	27 g	0.6%
Mustard powder	2¼ tsp	0.5 oz	15 g	0.4%
Fennel seeds, toasted and ground	2 tsp	0.5 oz	15 g	0.4%
Caraway seeds, toasted and ground	2 tsp	0.5 oz	15 g	0.4%
Maple sugar or brown sugar	1½ tsp	0.5 oz	15 g	0.4%
Coriander seeds, toasted and ground	2 tsp	0.5 oz	15 g	0.4%
Paprika	4 tbsp	0.9 oz	25 g	0.6%
Ground cayenne pepper	1 tsp	0.2 oz	4 g	0.2%
Freshly ground black pepper	1 tsp	0.4 oz	10 g	0.2%
Onion powder	1 tbsp	0.7 oz	20 g	0.5%
Garlic powder	1 tbsp	0.7 oz	20 g	0.5%
Apple cider vinegar	¼ cup	1.5 oz	43 g	1%
Baking soda	1 tsp	0.2 oz	7 g	0.2%
Fine sea salt (for brushing skin)	1½ tsp	0.4 oz	10 g	0.2%
Water	½ cup	3 oz	85 g	2%
Olive oil	as needed			

1. Butterfly the picnic shoulder. Turn it skin-side up and, using a stiff wire brush or fork, scrape and puncture the skin to create small holes and a rough surface. (You can use an ice pick, too.)

2. Combine the salt (for seasoning rub), mustard powder, fennel seeds, caraway seeds, maple sugar, coriander seeds, paprika, cayenne, black pepper, onion powder, garlic powder, and cider vinegar. Flip the roast back over and rub the mix into all the crevices on the meat side of the roast. Tie up the roast very tightly, keeping all the rub inside the roast.

3. Dissolve the baking soda and salt (for brushing skin) in the water. Brush all of the baking soda mixture over the skin. Put the roast on a rack and let stand, uncovered, in the refrigerator for 24 hours, or until the skin is very dry.

4. After 24 hours, lightly score the shoulder, being careful not to penetrate the meat, only the skin.

5. Preheat the oven to 500°F/260°C/gas 10. Place a rack in a roasting pan and set the roast on top. Insert a probe thermometer into the center of the meat next to but not touching the bone. Rub the skin with olive oil and roast for 55 minutes, then rotate the pan and add the water to the pan. Turn down the oven to 215°F/100°C/gas $\frac{1}{8}$ and slow-roast the meat until the internal temperature reaches 150°F/66°C, about 5 hours more.

6. Allow the roast to rest, uncovered, for about 45 minutes, then snip and remove the strings. Slice with a serrated knife—so it will cut through the skin easily—and enjoy.

Smoked Pork Sirloin

SERVES 4

Belly is the more common cut for making bacon, but sirloin is nice and lean and fine textured, and brining helps keep all that moisture safely inside even when the meat is subjected to the dry heat of a smoker. Slice this super-smoky meat and fry up the slices for breakfast. If you want to, add a little extra fat to the pan, since (unlike belly bacon) the sirloin doesn't have much of its own to render out. I like to make bacon with sirloin because it's a great way to use this particular cut of meat—and I have *plenty* of other uses for the belly—plus you end up with a little mini ham.

Master Brine (page 16), completely cold	8⅓ cups	67 oz	1,900 g	28.7%
Boneless pork sirloin or cowboy "ham" steak (see page 224)	1 whole	27 oz	766 g	71.3%
Rendered pork fat for cooking (optional)	as needed			

1. In a nonreactive container, brine the sirloin, fully submerged, in your refrigerator for 24 hours. Rinse well under cold water.

2. Prepare a smoker with about 2 cups/8 ounces of apple or hickory wood chips. Insert a probe thermometer into the center of the sirloin and smoke the meat, ideally at about 230°F/110°C, until the internal temperature at the center reaches 150°F/65°C. (The smoke will peter out after a while; don't add more chips, or the meat will be *too* smoky.)

3. Let the meat cool, then refrigerate until ready to serve. Cut into thick slices and fry until crispy and golden, adding a little rendered pork fat to the pan, if you like. Enjoy for breakfast (or anytime of day).

Resources

Information, Tools, and Community

Excellent tool for exploring the anatomy of the pig:
porcine.unl.edu/porcine 2005/pages/index. jsp?what=rotation3d

Excellent tool for exploring the anatomy of the steer:
bovine.unl.edu/bovine3D/ eng/rota.jsp

A good resource for buying butchery tools and sausage equipment:
www.butcher-packer.com

More wonderful sources for staying informed and involved in your community:
www.civileats.com
www.meatpaper.com
www.thebutchersguild.com
www.proteinuniversity.com
www.eatwild.com
www.eatwellguide.org

Great Butchers and Butcher Shops

WEST
Avedano's Holly Park Market, in San Francisco
www.avedanos.com

Bi-Rite Market, in San Francisco
www.biritemarket.com

Fatted Calf, in Napa Valley and San Francisco
www.fattedcalf.com

Laurelhurst Market, in Portland, Oregon
www.laurelhurstmarket.com

Lindy & Grundy, in Los Angeles
www.lindyandgrundy.com

Marina Meats, in San Francisco
www.marinameats.com

Olympic Provisions, in Portland, Oregon
www.olympicprovisions.com

Rain Shadow Meats, in Seattle
rainshadowmeats.com

EAST
Dickson's Farmstand Meats, in New York City
dicksonsfarmstand.com

Fleisher's, in upstate New York
www.fleishers.com

Larry's Custom Meats, in Hartwick, New York
(607) 293-7927

Lobel's, in New York City
www.lobels.com

The Meat Hook, in Brooklyn, New York
www.the-meathook.com

SOUTH
Belmont Butchery, in Richmond, Virginia
belmontbutchery.com

Cochon Butcher, in New Orleans
www.cochonbutcher.com

Dai Due Butcher Shop, in Austin
daidueaustin.com

MIDWEST/CENTRAL
Clancey's Meats, in Minneapolis
www.clanceysmeats.com

Fox & Obel, in Chicago
www.fox-obel.com/butcher-and-seafood-shop

Goose the Market, in Indianapolis
www.goosethemarket.com

Tony's Market, in Denver
www.tonysmarket.com

MONTREAL
Jean-Talon Market
www.marche-jean-talon.com

Porc Meilleur
www.marche-jean-talon.com

Resources

TORONTO
Cumbrae's
www.cumbraes.com

The Healthy Butcher
www.thehealthybutcher.com

VANCOUVER
Armando's Finest Quality
Meats
www.armandosmeats.com

The Butcher
thebutcher.ca/lamb.html

LONDON
The Ginger Pig
www.thegingerpig.co.uk

C. LIDGATE
www.lidgates.com

M Moen & Sons
www.moen.co.uk

A Few of My Favorite Farmers and Ranchers

WEST
Hudson Ranch
www.hudsonia.com

Ingel-Haven Ranch
and Magruder Meats
www.magrudergrassfed.com

Mariquita Farm
www.mariquita.com

Rocky Mountain Wooly
Weeders/Napa Valley
Lamb Company
woolyweeders.com

Sweet Briar Farms
sweet-briar-farms.com

EAST
EcoFriendly Foods
www.ecofriendly.com

Four Story Hill Farm
Honesdale, PA 18431
570-224-4137
pryzant@ezaccess.net

Moon in the Pond Farm
www.mooninthepond.com

SOUTH
Newman Farm
www.newmanfarm.com

Polyface, Inc.
www.polyfacefarms.com

MIDWEST/CENTRAL
Becker Lane Organic Farm
www.beckerlaneorganic.com

Grass Run Farm
www.grassrunfarm.com

Lazy S Farms
www.lazysfarms-glasco.com

Viking Lamb LLC
vikinglamb.com

Index

Chronicle Books publishes distinctive books and gifts. From award-winning children's titles, best-selling cookbooks, and eclectic pop culture to acclaimed works of art and design, stationery, and journals, we craft publishing that's instantly recognizable for its spirit and creativity. Enjoy our publishing and become part of our community at www.chroniclebooks.com.